HOW TO WRITE & PRESENT

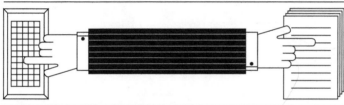

TECHNICAL INFORMATION

 2 EDITION BY CHARLES H. SIDES

ORYX PRESS
1991

HOW TO
WRITE &
PRESENT

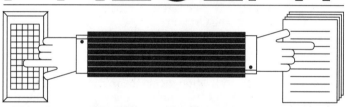

TECHNICAL
INFORMATION

 2 EDITION BY CHARLES H. SIDES ORYX PRESS 1991

FIRST EDITION TITLED:

How to Write Papers and Reports about Computer Technology

The rare Arabian Oryx is believed to have inspired the myth of the unicorn. This desert antelope became virtually extinct in the early 1960s. At that time several groups of international conservationists arranged to have 9 animals sent to the Phoenix Zoo to be the nucleus of a captive breeding herd. Today the Oryx population is nearly 800, and over 400 have been returned to reserves in the Middle East.

Copyright © 1991 by The Oryx Press
4041 North Central at Indian School Road
Phoenix, Arizona 85012-3397

Published simultaneously in Canada

Printed and Bound in the United States of America

∞ The paper used in this publication meets the minimum requirements of American National Standard for Information Science—Permanence of Paper for Printed Library Materials, ANSI Z39.48, 1984.

Library of Congress Cataloging-in-Publication Data
Sides, Charles H., 1952–
    How to write and present technical information / by Charles H. Sides. — 2nd ed.
    p. cm.
    Rev. ed. of: How to write papers and reports about computer technology. c1984.
    Includes bibliographical references and index.
    ISBN 0-89774-627-9
    1. Technical writing. 2. Communication of technical information. I. Sides, Charles H., 1952– How to write papers and reports about computer technology. II. Title.
T11.S528 1991
808'.0666—dc20   91-23244
                                                                          CIP

For Adam and Hannah
May you grow in wisdom and strength

Table of Contents

Preface

Technical writing is not reality but a re-creation of reality to meet the needs of readers. In 1983 I identified that as the theme of the first edition of *How to Write Papers and Reports about Computer Technology*. It remains one of the major themes of this second edition, now titled *How to Write and Present Technical Information*. Readers, particularly in high-tech industries, are no better suited today than nearly a decade ago for a minutely detailed account in a report or paper. They still have little interest in or time for poorly written, rambling accounts of technical information. They need to know what the main points of a communication are, why these points are important, and what they—the readers—should do with them.

A second, and almost as important, theme of this second edition is that not all of us read, write, and work equally well in the same fashion. We each have a style that suits us, and one of the goals of this book is to help individuals understand their work styles and the roles they play in written communication. For example, some people are "note writers." They write down every scrap of information they encounter when researching a documentation project. Everything is neat, organized, and easily accessible. Other people (myself included) are "head writers." These people do not write down much of the information they research but mull it over in their heads, organizing and reorganizing (often for weeks) before they put word to paper. It's pretty easy to tell when someone whose style resembles the first example is working. There is tangible evidence of it. But it's much more difficult to tell in the second example. These people appear to be daydreaming or goofing off.

Remember, however, *both styles work!* And if you are comfortable with one or the other, don't let someone talk you out of it simply because it is not his or her style. In other words, watch out for people who have the "do-it-right, do-it-my-way" syndrome. They're dangerous, and they might cause you to do less than your best work. For example, for both editions of this book, I was given between six and nine months to produce a draft. In each case I spent all but the last six weeks thinking about the subject, planning it, and organizing it—in my head. It's now approximately six weeks before the draft is due. Will I finish? You bet!! Will I be pleased with what I have done? Absolutely!! This style works for me, and it may work for you. If so, use it. If not, don't. We will look at the issue of work styles more in Chapter 1 of this book.

While this book is still intended for people who work with various technologies, it is not limited by that condition. What the book now contains can be used equally well in a variety of fields. Basically, it offers commonsense advice on how to write reports and papers that do not fail. The book could be used in a college technical writing course; the first edition was so used regularly. But neither the first edition nor this second edition is specially aimed at that audience. It is primarily for you, engineers, managers, developers, and technical writers, who have to write documents as quickly and as effectively as you can.

Several years back, *How to Write Papers and Reports about Computer Technology* was referred to as "the most product-oriented book on technical writing on the market." I took that as a compliment—even though the statement was, I'm sure, not intended to be. Current technical writing pedagogy has borrowed too heavily from the process approaches to teaching freshman composition. These approaches are excellent educational tools for students who are just entering college, many of whom have less than rudimentary writing skills. But the process approach to writing is best taught in the primary grades, thereby preventing large categories of writing problems before students arrive at a college. The approach does not work well for people who have to write a lot with very little time in which to do it—in other words, for professionals like you. Some applications of the process approach require you to write, write, and write some more until (by the very act of writing) you figure out what you have to say. It's not far from the belief that if you put enough monkeys armed with pens in a sealed room eventually one of them will write *Hamlet*. You do not have the time to write in order to discover your purpose, regardless of how valuable an educational exercise that might be. The result would be that almost everything you wrote before your purpose came to you would have to be discarded as useless.

A third major theme of this book, then, is that you are producing a product. It has become trendy to refer to documents about high technology as information products, and that makes sense. In this book, you will find out how to design, organize, and write them.

Introduction

SCENARIO 1

A short while ago, you were hired by an internationally known electronics firm that is planning to enter the highly competitive office automation market with a new, powerful 32-bit microcomputer. As a software engineer fresh out of college, you were hired to work in the Software Systems Laboratory, designing the new Unix-based operating system. Initially thrilled at such an opportunity and challenge, you have found out that as much as 40% of your time is spent communicating to others what you are doing—writing specifications, weekly status reports, making presentations in company meetings, interviewing people in marketing, and talking on the phone. Not only is this not what you were led to expect while you were in college, but also you realize that Freshman English, which you struggled through and frankly hated, did nothing to prepare you for the rigors of on-the-job communication.

SCENARIO 2

You have worked in the computer industry since the early days, the midfifties to be exact—a time when "bug in the system" meant looking for Raid or a flyswatter. You've seen a lot of changes, a lot of people who have come and gone, tremendous advances in the field, and periods of economic boom and bust. Your company has just gone through one of the bust periods, and a new boss has been hired to get things back on track. As usual in management changes, he is shaking things up, putting his own stamp on company operations. But for you there's a difference: He's told you that your written reports are unsatisfactory and that *they will be improved*. You realize that riding out your career to retirement will increasingly depend on your ability to satisfy whatever he means by clear, usable communication.

SCENARIO 3

After months of hard work, long hours, and sleepless nights, you have perfected your company's advance in 486-based PCs. The product will be shipped next month, and the advertising people are already touting it in the trade magazines. The division manager has selected you, naturally, as the best person to write an article for one of these magazines. Of course, you're honored. But you don't know the first thing about writing a journal article, never having thought of yourself as much of a writer.

SCENARIO 4

Your company has decided to send you to the industry's big, annual, international conference in Singapore next month. You will be expected to make a formal presentation before an audience of at least 150 top industry professionals from around the world. Even giving short presentations at small, informal group discussions makes you nervous. The thought of standing up in front of a large crowd in an auditorium is terrifying.

Each of these scenarios is common to professionals in high-tech industry. Many people, however, find themselves unprepared to meet the challenges of these communication tasks. They spend far more time than is necessary preparing for (and fretting about) these duties, and they get far poorer results than all this time should earn.

If you have ever found yourself in one of these scenarios, then this book is intended for you. Even if you don't recognize any of these situations now, you will eventually encounter these communication tasks if you intend to advance in high-tech industries. This book is for you, too.

PART 1

Writer, Audience, and Documentation

CHAPTER 1

Who We Are and What We Do

Who was Carl G. Jung, and what does his theory of personality types have to do with writing papers and reports in the high-tech industries? The first question can be easily answered: Jung was one of the most influential psychologists of the twentieth century, an influential thinker and prolific writer who with his theory of personalities explored psychological wellness, a largely ignored area of psychology at the time. The answer to the second question is summed up by this chapter's title, "Who We Are and What We Do," and it will be answered over the course of the next few pages. In short, who we are greatly affects what we do, how we do it, and why.

"DIFFERENT STROKES . . ."

Jung's theory of personality types suggests that individuals' personalities are best described by how two attitudes (extraversion and introversion) and four functions (sensation, intuition, feeling, and thinking) interrelate. The psychological jargon is unimportant here; what is important is how these things work.

Extraversion (Jung's spelling) describes individuals who prefer to focus their lives outward to the experiences of other people and things. They tend to be active and energetic—doers. Introversion describes individuals who prefer to focus their lives inward to the experiences of thought and reflection. They are no less active than extraverts, but their activity is internalized and hidden from the view of others. Realize that society has associated these terms with some negative inferences. Extraversion does not refer to blabber-mouthed boors any more than introversion refers to reclusive hermits. In fact, Jung originally coined these words to mean exactly what their roots suggest: "extra-vert"—turning out, and "intro-vert"—turning in.

Sensing and intuition are perceptive functions, again more psychological jargon to describe how people take in (perceive) information. Those who prefer sensing would rather rely on their five senses alone. Those who prefer intuition use their five senses only to gain enough information to make an educated guess or hunch about what they have perceived.

Both sensing and intuition provide the information with which we make judgments. And we do that either through feeling or thinking. People who prefer "feeling judgment" generally fit their perceptions into some preconceived value system (cultural, personal, corporate, familial, religious) before deciding what to do with the information. People who prefer "thinking judgment" usually decide upon information based solely on its merit, irrespective of fitting into a value system. In this case, too, society has colored the issue with negative connotations. Feeling does not mean weak and emotional, and thinking does not mean dispassionate and logical. In fact, Jung thought of both as rational, trustworthy, and valuable ways of making judgments.

Although this explanation is grossly oversimplified, it is about as far as Jung explored the issue explicitly in his 1923 book, *Psychological Types*. He did, however, imply two other attitudes—judgment and perception—which Katherine Briggs and her daughter Isabel Briggs Myers described fully in their later work, which led to a psychological evaluation instrument (the Myers-Briggs Type Indicator, or MBTI for short). Essentially, judgment refers to people who would prefer bringing matters to closure. Perception refers to people who would prefer to keep matters open-ended. The MBTI enables people, with the aid of trained professionals, to determine their type preferences, bettering their understanding of themselves and of others.

So, again, what does this have to do with us? Simple: four two-part preferences yield 16 personality types, which are described in the following chart. (See Table 1.)

Understand, and this is extremely important, that these personality types are not stereotypes into which you can be pigeon-holed. In fact, Jung contended that each one of us has some aspect of every single category. It's just that we *prefer* to use some categories over others. A helpful comparison

Table 1: Type Table

	SENSING TYPES		INTUITIVES	
	WITH THINKING	WITH FEELING	WITH FEELING	WITH THINKING

	ISTJ	**ISFJ**	**INFJ**	**INTJ**
INTROVERTS — JUDGING	Serious, quiet, earn success by concentration and thoroughness. Practical, orderly, matter-of-fact, logical, realistic and dependable. See to it that everything is well organized. Take responsibility. Make up their own minds as to what should be accomplished and work toward it steadily, regardless of protests or distractions. Live their outer life more with thinking, inner more with sensing.	Quiet, friendly, responsible and conscientious. Work devotedly to meet their obligations. Lend stability to any project or group. Thorough, painstaking, accurate. May need time to master technical subjects, as their interests are not often technical. Patient with detail and routine. Loyal, considerate, concerned with how other people feel. Live their outer life more with feeling, inner more with sensing.	Succeed by perseverance, originality and desire to do whatever is needed or wanted. Put their best efforts into their work. Quietly forceful, conscientious, concerned for others. Respected for their firm principles. Likely to be honored and followed for their clear convictions as to how best to serve the common good. Live their outer life more with feeling, inner more with intuition.	Have original minds and great drive which they use only for their own purposes. In fields that appeal to them they have a fine power to organize a job and carry it through with or without help. Skeptical, critical, independent, determined, often stubborn. Must learn to yield less important points in order to win the most important. Live their outer life more with thinking inner more with intuition.
	ISTP	**ISFP**	**INFP**	**INTP**
INTROVERTS — PERCEPTIVE	Cool onlookers, quiet, reserved, observing and analyzing life with detached curiosity and unexpected flashes of original humor. Usually interested in impersonal principles, cause and effect, or how and why mechanical things work. Exert themselves no more than they think necessary, because any waste of energy would be inefficient. Live their outer life more with sensing, inner more with thinking.	Retiring, quietly friendly, sensitive, modest about their abilities. Shun disagreements, do not force their opinions or values on others. Usually do not care to lead but are often loyal followers. May be rather relaxed about assignments or getting things done, because they enjoy the present moment and do not want to spoil it by undue haste or exertion. Live their outer life more with sensing, inner more with feeling.	Full of enthusiasms and loyalties, but seldom talk of these until they know you well. Care about learning, ideas, language, own. Apt to be on yearbook staff, perhaps as editor. Tend to undertake too much, then somehow get it done. Friendly, but often too absorbed in what they are doing to be sociable or notice much. Live their outer life more with intuition, inner more with feeling.	Quiet, reserved, impersonal. Enjoy especially in theoretical or scientific subjects. Logical to the point of hair-splitting. Interested mainly in ideas, with little liking for parties or small talk. Tend to have very sharply defined interests. Need to choose careers where some strong interest of theirs can be used and useful. Live their outer life more with intuition, inner more with thinking.
	ESTP	**ESFP**	**ENFP**	**ENTP**
EXTRAVERTS — PERCEPTIVE	Matter-of-fact, do not worry or hurry, enjoy whatever comes along. Tend to like mechanical things and sports, with friends on the side. May be a bit blunt or insensitive. Adaptable, tolerant generally conservative in values. Dislike long explanations. Are best with real things that can be worked, handled, taken apart or put back together. Live their outer life more with sensing, inner more with thinking.	Outgoing, easygoing, accepting, friendly, fond of a good time. Like sports and making things. Know what's going on and join in eagerly. Find remembering facts easier than mastering theories. Are best in situations that need sound common sense and practical ability with people as well as with things. Live their outer life more with sensing, inner more with feeling.	Warmly enthusiastic, high-spirited, ingenious, imaginative. Able to do almost anything that interests them. Quick with a solution for any difficulty and ready to help anyone with a problem. Often rely on their ability to improvise instead of preparing in advance. Can always find compelling reasons for whatever they want. Live their outer life more with intuition, inner more with feeling.	Quick, ingenious, good at many things. Stimulating company, alert and outspoken, argue for fun on either side of a question. Resourceful in solving new and challenging problems, but may neglect routine assignments. Turn to one new interest after another. Can always find logical reasons for whatever they want. Live their outer life more with intuition, inner more with thinking.
	ESTJ	**ESFJ**	**ENFJ**	**ENTJ**
EXTRAVERTS — JUDGING	Practical realists, matter-of-fact, with a natural head for business or mechanics. Not interested in subjects they see no use for, but can apply themselves when necessary. Like to organize and run activities. Tend to run things well, especially if they remember to consider other people's feelings and points of view when making their decisions. Live their outer life more with thinking, inner more with sensing.	Warm-hearted, talkative, popular, conscientious, born cooperators, active committee members. Always doing something nice for someone. Work best with plenty of encouragement and praise. Little interest in abstract thinking or technical subjects. Main interest in things that directly and visibly affect people's lives. Live their outer life more with feeling, inner more with sensing.	Responsive and responsible. Feel real concern for what others think and want, and try to handle things with due regard for other people's feelings. Can present a proposal or lead a group discussion with ease and tact. Sociable, popular, sympathetic. Responsive to praise and criticism. Live their outer life more with feeling, inner more with intuition.	Hearty, frank decisive, leaders in activities. Usually good in anything that requires reasoning and intelligent talk, such as public speaking. Are well-informed and keep adding to their fund of knowledge. May sometimes be more positive and confident than their experience in an area warrants. Live their outer life more with thinking, inner more with intuition.

is handedness. Most of us are either right- or left-handed, but all of us can use the unpreferred hand. Personality type is the same. Some of us are introverted, but we are able to interact with other people on a social and professional basis. Some of us prefer thinking judgment, but we are able to use and understand value systems.

PERSONALITY TYPE AND WORK

Much technical writing is done by teams of people, each bringing different work habits and preferences to their work. (See Table 2.) Understanding those differences enables you to be a more productive team member, appreciating why you and others around you prefer to do what you do. It's common sense that you should work better doing what you prefer in a way that you prefer it. That sounds ridiculously simple, but since most people think "my way is the right way," it's not as easy as it may seem. A good deal of trust is required to allow your colleagues and employees to work in a way that is productive for them but counter-productive for you. If you can do it, the results will astound you.

CONCLUSION

In this chapter, we have briefly considered the effects that an individual's personality can have on his or her work. This material forms an underlying current for much of the rest of this book, from audience analysis to report design. In short, remember that we do not all think the same way, behave the same way, value the same things, or work the same way. It's a tall order, but we need to try to make the information we communicate in reports and papers accessible to as many different types of people as possible. Realize that this means we do not all *read* the same way, either. Some read every word meticulously. Others skim. One of the goals of this book is to show you ways to design reports so that they succeed for both reading strategies.

SUGGESTED READINGS

Jung, Carl G. *Psychological Types*. Princeton, NJ: Princeton University Press, 1971.

Kiersey, David and Marilyn Bates. *Please Understand Me: Character and Temperament Types*. Del Mar, CA: Prometheus Nemesis Books, 1978.

Lawrence, Gordon. *People Types and Tiger Stripes*. Gainesville, FL: CAPT, Inc., 1983.

Myers, Isabel Briggs. *Gifts Differing*. Palo Alto, CA: Consulting Psychologists Press, 1980.
Sides, Charles H. "What Does Jung Have to Do with Technical Writing?" *Technical Communication*. Vol. 36, No. 2 (1989), pp. 119-26.

Table 2: Personality Type and Work

EXTRAVERTS	INTROVERTS
Like variety and action.	Like quiet for concentration.
Tend to be faster, dislike complicated procedures.	Tend to be careful with details, dislike sweeping statements.
Are often good at greeting people.	Have trouble remembering names and faces.
Are often impatient with long slow jobs.	Tend not to mind working on one project for a long time uninterruptedly.
Are interested in the results of their job, in getting it done and in how other people do it.	Are interested in the idea behind their job.
Often do not mind the interruption of answering the telephone.	Dislike telephone intrusions and interruptions.
Often act quickly, sometimes without thinking.	Like to think a lot before they act, sometimes without acting.
Like to have people around.	Work contentedly alone.
Usually communicate freely.	Have some problems communicating.

SENSING TYPES	INTUITIVE TYPES
Dislike new problems unless there are standard ways to solve them.	Like solving new problems.
Like an established way of doing things.	Dislike doing the same thing repeatedly.
Enjoy using skills already learned more than learning new ones.	Enjoy learning a new skill more than using it.
Work more steadily, with realistic idea of how long it will take.	Work in bursts of energy powered by enthusiasm, with slack periods in between.
Usually reach a conclusion step by step.	Reach a conclusion quickly.
Are patient with routine details.	Are impatient with routine details.
Are impatient when the details get complicated.	Are patient with complicated situations.
Are not often inspired, and rarely trust the inspiration when they are.	Follow their inspirations, good or bad.
Seldom make errors of fact.	Frequently make errors of fact.
Tend to be good at precise work.	Dislike taking time for precision.

THINKING TYPES	FEELING TYPES
Do not show emotion readily and are often uncomfortable dealing with people's feelings.	Tend to be very aware of other people and their feelings.
May hurt people's feelings without knowing it.	Enjoy pleasing people, even in unimportant things.
Like analysis and putting things into logical order. Can get along without harmony.	Like harmony. Efficiency may be badly disturbed by office feuds.
Tend to decide impersonally, sometimes paying insufficient attention to people's wishes.	Often let decisions be influenced by their own or other people's personal likes and wishes.
Need to be treated fairly.	Need occasional praise.
Are able to reprimand people or fire them when necessary.	Dislike telling people unpleasant things.
Are more analytically oriented—respond more easily to people's thoughts.	Are more people-oriented—respond more easily to people's values.
Tend to be firm-minded.	Tend to be sympathetic.

JUDGING TYPES	PERCEPTIVE TYPES
Work best when they can plan their work and follow the plan.	Adapt well to changing situations.
Like to get things settled and finished.	Do not mind leaving things open for alterations.
May decide things too quickly.	May have trouble making decisions.
May dislike to interrupt the project they are on for a more urgent one.	May start too many projects and have difficulty in finishing them.
May not notice new things that need to be done.	May postpone unpleasant jobs.
Want only the essentials needed to begin their work.	Want to know all about a new job.
Tend to be satisfied once they reach a judgment on a thing, situation, or person.	Tend to be curious and welcome new light on a thing, situation, or person.

CHAPTER 2

What Is High Quality Documentation?

Better yet, what is documentation? That this basic question is necessary and the number of answers it receives show the problems that exist in writing about technology. Is documentation the manuals alone? The help screens? The memos that communicate information on a daily basis and lead to reports and manuals? The reports themselves? How about papers and articles that are published in trade magazines? Small group presentations? Formal presentations? All of the above? The last answer is the correct one. Documentation, when accurately defined, embraces all acts of writing and speaking about high technology to all possible audiences. It is an enormous communication task, fraught with snares and minefields. This complexity is perhaps the reason why many people identify poor technical writing solely with the computer industry. Too often, they are right.

CHARACTERISTICS OF GOOD DOCUMENTATION

Now that we have decided what documentation is, we can move on to the more important problem of identifying the characteristics of good documentation. They do exist.

Meeting the Audience's Needs

Good documentation satisfies the needs of an identified audience. This concept is the most important aspect of writing in the high-tech industries. The next chapter is devoted entirely to the subject of finding out who your audience is and what they want to read. Readers come to a document—be it memo, report, manual, or paper—with certain needs. They might be mildly interested in a new development in the field; they might need to get a "go-ahead" before proceeding on a new project; they might want to see if a project can be completed economically or if it has a place in the projected market; or they might want to learn a programming procedure in as little time as possible. Whatever the readers' needs, they must be met by the document. If they are not, the importance of the subject, the brilliance of the author, will be irrelevant. The communication will fail. Remember, technological advancement is useless without the ability to communicate it to customers.

Good Organization

Once the author has considered the audience, organization is the second most important concern. Good documentation is rigorously organized. In fact, organization is one of the ways audience needs can be met. But as we have seen already, different types of audiences might prefer different organizational strategies. Providing multiple pathways to information is a way to ensure that documents are useful to broad ranges of readers.

Documentation is writing to be used. One rarely thinks of somebody curling up in front of a cozy fire on a cold winter evening with a user manual. Documentation does not, therefore, have to pique the interests of readers, because readers already come to documentation with specific needs. Readers *are* interested; otherwise they would not have picked the document up in the first place. This does not mean, however, that documentation writers need not be concerned with readers' interests. Everything in a document must be organized with the reader in mind. The document must present the subject in such a way that all readers can get "from here to there," from being interested in the topic to knowing enough about the topic to do their jobs. And readers had better be able to do this without false steps, circumlocutions, and journeys down tangential paths. One of the best ways to ensure that readers will put your document down without finishing it is to confuse them.

Now, I just said that a document does not have to stir interest; but that does not mean it has to be boring. Technology is fascinating, and the only thing that could possibly be boring about it is how writers treat it in documents. Nor does this mean that documents cannot be persuasive. Subtle persuasion is a handy tool for high-tech professionals, and it can be achieved through the clear explanation of a topic in which readers are already interested. In a way, this assumed interest makes people who write about high tech lucky. They don't have to work as hard to engage their readers. But certain stylistic techniques can be used to further enhance the topic: a varied sentence structure, variable sentence lengths, a personal style, humor.

Humor

Humor!? What about humor? Is there a place for it in this type of technical communication? Not using humor in technical writing is similar to the school marm's admonition against starting sentences with *and*, and

to warn *(n.)*
gently
admonish *(v.)*

every knowledgeable eighth grader's realization that Stephen King does it all the time. The question is moot. Good documentation (particularly manuals and magazine articles, not so much reports and memos) is rife with subtle humor—much of it intentional. As a result, documentation writers must be the funniest people in high-tech today. Seriously, their humor, when done right, never gets in the way of the subject. And for certain audiences, it makes a difficult subject easier to understand.

Jargon

Good documentation uses jargon. That's right. Good documentation uses jargon. But that's heresy, isn't it—a sell-out to those technocrats and bureaucrats who would wreck our language? Taken by itself, the statement about jargon is not entirely accurate, but it's not heresy either. The computer field and other high-tech industries are extremely complex. Part of the reason for this complexity is that, compared to medicine, astronomy, and even quantum mechanics, the high-tech field is relatively new. A large part of the documentation produced in this field is written for other people in the industry, readers who have considerable knowledge of the subject. This situation introduces the issue of audience, and that's the proper context in which to view the use of jargon. Will the readers share the writer's knowledge of terminology? If so, altering that terminology makes no sense. At worst, doing so oversimplifies the topic, damages its integrity, and limits the imagination of the writer. But, and it's an important *but*, authors must have a clear idea of who the audience is. Even managers in high-tech industries may not share the technical vocabulary of their engineers. In fact, some engineers do not always share the same vocabulary with other engineers. Consequently, the use of jargon is audience-dependent. Remember: Always use what the audience will understand.

To paraphrase Samuel Taylor Coleridge, whatever can be written in fewer and simpler words without damaging the audience's understanding of the information is poorly written. There is no better guideline for the use of jargon than that.

Readability

Readability (or usability) is the final characteristic of good documentation. Readability is the skeleton of all writing that works. We can't see it, but it holds the writing up and allows it to do things. Readability is achieved in a number of ways. Sentence and word length have something to do with it. So does sentence structure. Most readability formulas, such as Gunning's Fog Index and the Flesch Test, are based on these criteria. But these tests oversimplify the issue. Correctly structured, with a feel for rhythm and the ebb-and-flow of a phrase, even the long sentence is readable. When it is not overused, it can be a way to attain structural variety, or to emphasize weighty importance. Rather than rely on artificial means for determining readability, realize this: The more often you make a reader reread to understand what you have said, the less readable (and usable) your document is.

CONCLUSION

In this chapter I have identified several basic characteristics of good documentation. Notice that I did not mention formats, except as examples in passing. Good documentation is independent of the format that contains it. At its most basic level, good documentation is good writing and good speaking. It is also good sense.

To that end, the rest of this book will examine in detail the techniques of good writing and speaking as they can be applied to information products within the high-tech industries.

SUGGESTED READINGS

Arzberger, C. R. "Making a Long Story Short—in Common Language." *Bell Laboratories Record*. Vol. 59 (1981), pp. 305-10.

Benzon, William L. "The Computer and Technical Communication." *Journal of Technical Writing and Communication*. Vol. 11, No. 2 (1981), pp. 103-08.

Brockmann, R. John. *Writing Better Computer Documentation*. New York: Wiley-Interscience, 1986.

Browning, Christine. *Guide to Effective Software Technical Writing*. Englewood Cliffs, NJ: Prentice-Hall, 1983.

Casey, Bernice E. "The Impact of the Technical Communicator on Software Requirements." *Journal of Technical Writing and Communication*. Vol. 11, No. 4 (1981), pp. 361-72.

Felker, Daniel B., ed. *Document Design: A Review of the Relevant Research*. Washington, DC: American Institute for Research, 1980.

—————. *Guidelines for Document Designers*. Washington, DC: The Document Design Center, 1981.

Grimm, Susan J. "EDP User Documentation: The Missing Link." *IEEE Transactions on Professional Communication*. Vol. PC-24 (1981), pp. 79-83.

Harper, John. *Data Processing Documentation: Standards, Procedures, and Applications*. Englewood Cliffs, NJ: Prentice-Hall, 1976.

Holden, F. W. "A System for Documenting Computer Programs." *20th International Technical Communication Conference Proceedings*. Washington, DC: Society for Technical Communication, 1973, pp. 101-05.

Petito, Joseph. "Computer Software Documentation." *Journal of Technical Writing and Communication*. Vol. 4, No. 1 (1974), pp. 117-20.

Sides, Charles H. *Technical and Business Communication: Bibliographic Essays for Teachers and Corporate Trainers*. Urbana, IL: NCTE, 1989.

——————. "Writing in the Computer Industry." *Journal of Technical Writing and Communication*. Vol. 16, No. 2 (1986), pp. 1-7.

CHAPTER 3

How to Define Your Audience

The first question that people who write about high-tech products might want to ask is, "Why is audience definition so important?" The answer is simple: audience definition helps writers specifically target the major group of readers for a document. It enables writers to discover what their readers know about the subject of a report or paper. It focuses on what the readers need to know in order to perform their jobs better or in order to increase their knowledge about the subject. And it helps writers determine what their readers will do with the information they read. The lack of conscious audience definition is responsible for the regrettable state of much technical writing: documents with poorly defined purposes, written for readers whom the writer has not considered and does not understand.

Viewed in the proper context, audience definition is the single most important aspect of rendering technical information usable to readers. This is the underlying principle of this book. The system presented in this chapter will provide readers with an easy-to-use method for defining the audience.

THE AUDIENCE DEFINITION SYSTEM

A systematic method for audience definition is divided into four processes: (1) defining who the readers are; (2) defining what the readers know; (3) defining what the readers need to know; and (4) defining what the readers will do with the information provided. Figure 1 depicts how these processes interact in helping to determine the message to be communicated.

Who Are the Readers?

If you are going to write a paper or a report about a high-tech subject, the first thing you need to know is who your readers are. You need to know this as specifically and in as much detail as possible. Because of the daily technological advances in high-tech industries, you are often writing in a vacuum without specific knowledge about your intended readers. Consider these examples. Are your readers managers in firms that use computers? Engineers in firms that develop hardware or software?

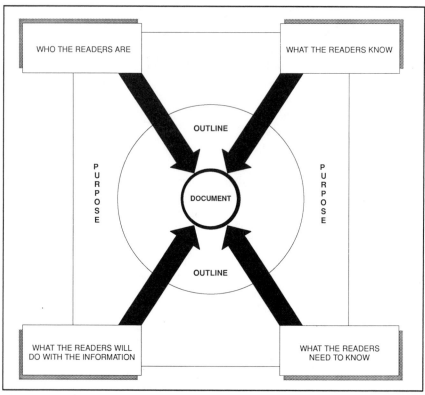

WHO THE READERS ARE

WHAT THE READERS KNOW

OUTLINE

PURPOSE

DOCUMENT

PURPOSE

OUTLINE

WHAT THE READERS WILL DO WITH THE INFORMATION

WHAT THE READERS NEED TO KNOW

Figure 1. Audience Definition System

Financial planners in such firms? Students? Computer-phobes? People who are purchasing their first PC for the home or office? Or worse, is your audience made up of many of these groups?

For example, ZetaCorp, a hypothetical computer company has devoted its entire work force to developing a new line of office information systems. The project develops mountains of information which has to be communicated across the company. This communication, however, is destroying the project schedule. Hardware engineers discover that software engineers do not use the same vocabulary (jargon), even though both groups have similar educational backgrounds. The engineers find out that their reports are regularly sent to managers who have only a general idea of recent technological developments. The reports are abstracted for marketing and sales people who are even less technologically informed. Everyone thinks someone else is screwing up, and the project falls farther and farther behind schedule. Sound familiar?

Readers come to a document with different backgrounds, different levels of knowledge about the subject, and different needs for the information that is provided. Management in firms that use computers as part of their daily routine might read articles in trade magazines for information that will enable them to use their present systems more efficiently. Or they might find information which will suggest that they need to update or replace their systems. This information is especially important for small- or medium-sized organizations since microcomputers can now outperform the minicomputers of only a few years ago. Management looks to the trade magazines for trends and information that will affect their business for years to come. The writer's duty is to furnish this information in a form readers can understand and use.

Engineers in hardware and software development firms must keep abreast of the rapid technological development in their fields, particularly since these developments signal competition. These readers make up a highly technical audience which needs highly technical information. Writers must meet the needs of these readers, also.

Financial managers in these firms do not have the same technical background as do their engineers, but they need just as much information—with enough explanation so they can grasp the importance of new developments. They are concerned about competitors' developments because these developments might affect their own company's future. Even rumor of development is important to these readers; they must keep ahead of trends in order to plan company marketing and development strategies. Technical fliers, in addition to trade articles, provide these people with information.

Students might even come to a report or paper that is available to the public. They might already know a considerable amount about the report's subject, or they might know virtually nothing. They come for learning and for information that will shape their careers. Articles in popular computer journals furnish them with a great deal of this information.

Computer-phobes might encounter your technical information, and they should be dealt with carefully. Every computer-phobe who searches out your company's technology is a potential customer.

People who are buying their first computer might also find their way into your documents. Should they be planned for? If their first purchase of a computer product is from your company, you want to make sure it isn't their last. The answer is yes; plan for them, too.

Of course, the worst situation is when projected audiences overlap. What can you do then? And how do you account for the different preferences readers have for how they comprehend and use information? It seems hopelessly complex, but it is not. In such situations, your writing strategy depends on how much weight each component of your overall audience carries. Is it more important to focus on decision-makers or end-users—on people who want step-by-step information or overviews?

The following questions will help you identify prospective audiences. The answers to them will provide you with information to begin making judgments about the type and quantity of information to include in reports or papers as well as how to arrange that information. Later, this information will be your foundation for more in-depth considerations of readers' needs.

Who is the audience? Try to answer this question in as much detail as you can. For in-house reports, this means identifying your readers by name and position. For papers submitted for publication, it means finding out from the editors which type of reader is considered "typical." Being able to do this sort of analysis helps bring writers out of the vacuum of trying to communicate to nameless, faceless readers. It is much easier to write if you have some definite ideas about whom you are writing for.

What is the educational background of your audience? With this question, you are trying to determine if the audience shares a common educational background with you. For example, how current is the reader's education? Is the reader's educational experience similar to yours? Ask yourself if you could participate in a discussion of the topic with the reader on an equal level. These questions help refine your reader's identity.

What is the professional background of your audience? Here you are trying to determine whether the audience shares similar experiences with you. This question is important because it enables you to determine the appropriate vocabulary to use in the document, how much background is required, and how much definition of terms and concepts the audience needs.

These audience identification questions are important because they enable writers to get a picture of what the projected audience is like. Writing to an audience we can visualize is far easier than writing to an unindentified audience. Beyond that, identifying our audience forces us to start thinking about how our information should be organized.

What Does Your Audience Know about the Subject?

The first answer to this question is that it's hard to tell—without asking the readers personally, that is. The best way to discover what an audience knows is to do precisely that—ask them. But this requires very small, closed audiences—the sort that might be on the receiving end of an interoffice memorandum. The majority of communication environments preclude this possibility. Writers and readers are separated by space and time; they cannot interface directly. What is needed in this situation is a system that is a little more systematic than guesswork.

Take the following example. In the Venix operating system for PDP-11 computers, there is a feature, bit-mapping, which simulates split instruction and data space (I & D space). Now, if you are a software or a hardware engineer, this explanation is as much as you need to know to understand that the Venix operating system enables PDP-11 computers to run faster than it can under other operating systems. If you are a manager in a large, urban hospital and are considering improving your computer's performance, you may not have the slightest understanding of bit-mapping and split I & D space. But if the writer tells you that this feature will speed up your organization's computer, you will be interested. You will also need more explanation—such as how much faster is faster, and at what costs both in terms of cash outlay and in terms of features gained and lost. If you are a financial planner or marketer for a competing software company, you will know what the market is looking for. You will need to know how this new operating system is an improvement in order to determine how your company can meet its challenge. If you are a student, you may know that Venix is a Unix derivative, but you might want to know more about the history and development of Venix. If you are a newcomer to computers, you may not know what operating systems are, what the PDP-11 is, not to mention what bit-mapping and split I & D space are.

Based on the answers to who the audience is, the following questions will help you determine what the audience knows. From the answers to these questions, you can begin to specify what you want the audience to know and what they need to know.

What does the audience know about the specific topic of your report or paper? This differs from the first question in that it focuses the possible answers. Even though an audience might have some background knowledge of your topic, they might need considerable orientation to the specific topic you are reporting to them. For example, they might know a good deal about computer architecture, but would need to be led clearly through a report or paper about the new type of buss that your firm has developed. Be careful not to over-assume the level of understanding of your audience. This is the single easiest mistake to make in technical writing and the one that renders more information unusable than any other mistake. Even though your readers might share common educational and professional backgrounds with you, they have not stood by you while you developed the information for a document. Nor can they read your mind. Don't make them try.

How much background is necessary? This question must inevitably follow the first two. The answer to this question allows you to start thinking about the introduction of a report or the lead of a paper. For example, readers who share your educational background and professional experience will be bothered by long-winded introductions that tell them what they already know. Some research suggests that they will either skip over the irrelevant material or not read the document at all, routing it to someone else. Certainly this reader reaction is not one of your goals as a writer. But the opposite of this situation is just as bothersome to another audience. Because as writers we tend to think our readers will automatically understand what we are trying to say, the tendency is to drop uninformed readers into the middle of specifics before we have told them the significance of the communication. This guarantees bewilderment.

These questions we have been considering are important because they begin to shape the information you have gathered about a topic of a report or paper. The process of audience analysis, incrementally focusing what you know about the people who will read your reports or papers, makes organizing information a simpler task.

What Does the Audience Need to Know?

This matter is subjective for it depends on your purposes as a communicator. Is your purpose solely to provide information, as objectively as possible? Or do you intend to persuade readers to a particular point of view, perhaps even to buy a particular product? The answers to these two questions determine what you want your audience to know. In fact, the question is often best stated, "What *should* your audience need to know?"

A lot depends on your own professional situation. Are you working for a company that expects you to write an apparently objective paper, but also to display the benefits of a product they are developing? Such is often the case with papers published in various computer trade magazines. Or are you planning to give a talk at a professional convention or trade show? Here the situation is even trickier. Your company will want you to reveal enough information so that the audience will understand the importance of late developments at your company, but you will not want to "give away the store."

The following set of questions will help you identify what your readers need to know.

What information does your audience need from your report or paper in order to do their jobs better? The answer to this question forces you to start the process of examining all the information you have gathered about your topic so that you can delete what is not necessary. Too many writers feel compelled to jam all their material into a report or paper. After all, they worked hard to get that material; why throw it away? Editors in the trade press will make sure that irrelevant information does not stay in your papers, but *you* must ensure that it is deleted from your reports.

Will the information that is included need a technical slant or a managerial slant? The answer to this question complements the answer to the questions that identify your audience. Even though you may explain your subject in ways that make it important to both management and technical readers, the report or paper should have a primary focus toward one group or the other.

Sometimes writers are compelled by situations to give readers information they do not know they want or need. Accurate audience analysis enables you to do this effectively when it is called for.

What Will Readers Do with the Information?

An important aspect of deciding what information you will include in reports or papers is determining what readers will do with it. Knowing this enables you to avoid including information no one needs and cluttering the report needlessly. The following questions will help you identify what readers are likely to do with the information you provide them.

Will readers use the information in your report or paper to perform professional tasks? The answer to this question helps you to decide if your report or paper should have a procedural organization. Will readers need clearly identified steps to follow, as in a manual?

Will readers use the information in your report or paper to increase their knowledge of the field? The answer to this question helps you to decide if the audience needs to see how the information in your report or paper fits into the literature of their field.

Only by visualizing what readers will do with the information you plan to give them can you be sure that it is the right information, accurately communicated.

What personality types will be reading your report or paper? Keep in mind that different personality preferences affect audience needs for information. If you are writing for one or two people whose preference you know, then you can gear your presentation to their type. Otherwise, you will have to try to include information for the main categories of types in your audience. For example, intuitives generally need just enough information to get started. Overviews are important for them to get the sense of what they will be reading, as is information that enables them to skim a document. Sensing types will want more detailed information, meeting their needs for step-by-step analyses or procedures. Extraverts will often use your information verbally, usually in meetings. Introverts may file it for use in other documents. Whatever the situation, you need to think about these differences and the roles they play in your writing.

CONCLUSION

This chapter has presented a system for taking general information or gut feelings about the reader and turning these into specific details that you can use to help organize information for a report or a paper. In a sense, the method is subversive; it treats the matter of audience definition from the stance of giving the audience what it wants. This is a variation of the "customer is always right" philosophy. But any writer who has had a writing course has probably figured out the importance of this philosophy somewhere during the course. We neglect to realize, however, that the same situation exists in professional communication. We might not always tell readers what they want to hear, but we should always give them what they need—and should want—to know.

The following decision logic table summarizes the processes presented in this chapter (see Figure 2). It does precisely what its name suggests: provides information with which decisions can be made logically. To use this one as a checklist for defining audiences, match the conditions that you believe are present for the document you intend to write (Y=Yes, N=No). In this table, six combinations of conditions are possible. Depending on how the conditions you have identified for your report match up with those listed in the table, follow the appropriate actions marked with Xs.

For example, if you decide that you are addressing a management audience only and that your purpose is only to inform (conditions common to many memos), you would focus the document on matters that affected budget and personnel matters, schedules, or decisive actions that need to be taken, including the specific facts and supporting data as necessary. Depending on the needs of the individuals you are addressing, however, you might decide that some design information is necessary (e.g., addressing a hardware development project manager). That is okay; trust your audience analysis. Your primary focus for the information will probably still be on matters that show how the design information fits into the overall project plan and schedule.

And remember, different people read and interpret information in different ways. Provide several paths of access to important information.

		1	2	3	4	5	6	
CONDITIONS	Management Audience	Y	Y	Y	Y	N	N	Else
	Technical Audience	Y	Y	N	N	Y	Y	
	Informative Purpose	Y	N	Y	N	Y	N	
	Persuasive Purpose	N	Y	N	Y	N	Y	
ACTIONS	Include Budgetary & Personnel Information	X	X	X	—	–	–	–
	Include Specific Design Information	X	X	–	–	X	X	–
	Include "Facts & Figures"	X	X	X	X	X	X	–
	Include Common Ground Orientation Between Reader and Writer	–	X	–	X	–	X	–
	Rethink Report	–	–	–	–	–	–	X

Figure 2. Decision Logic Table for Audience Definition

SUGGESTED READINGS

Caskey, Drew S. "Technical Writing: The Importance of Reader Interest." *Journal of Technical Writing and Communication.* Vol. 3, No. 3 (1973), pp. 217-21.

Cederborg, Gibson A. "Write Disciplined Reports that Captivate Your Readers." *Chemical Engineering.* Vol. 82 (July, 1975), pp. 98-100.

Cottrell, Beekman W. "The Unceasing Need for Audience Awareness." *23rd International Technical Communication Conference Proceedings.* Washington, DC: Society for Technical Communication, 1976, pp. 120-21.

Holland, V. Melissa, et. al. "How Can Technical Writers Write Effectively for Several Audiences at Once?" In *Solving Problems in Technical Writing,* Lynn Beene and Peter White, eds. New York: Oxford University Press, 1988, pp. 27-54.

Humphrey, Susan R. and Gerald I. Williamson. "Make Your Technical Reports 'People-Oriented.'" *The American Business Communication Association Bulletin.* Vol. 35, No. 4 (1972), pp. 27-32.

Mathes, J.C. and Dwight Stevenson. *Designing Technical Reports.* 2nd Edition. New York: Macmillan, 1991.

Pearsall, Thomas E. *Audience Analysis for Technical Writing.* Beverly Hills, CA: Glencoe Press, 1969.
Wilkins, Keith A. "Receiver-Bound Concept of Communications." *Journal of Technical Writing and Communication.* Vol. 4, No. 4 (1974), pp. 305-13.

PART 2

Getting Started

CHAPTER 4

How to Get Organized

Every time I teach a communications seminar, I am inevitably told the same thing by course participants: "The most difficult thing for me to do is to get started." This is true for almost everyone. I don't think I have met a single person, including people who make their living by writing, who have said that they can sit down anytime, anywhere, and start writing. The source of this problem is getting organized, and if it is your problem you are not unique. Everyone experiences the problem of getting organized.

For some of us, it means an almost debilitating situation in which we will put off starting a document, using any excuse which is handy. "I don't have time right now because I am busy," is the most popular reason for not writing. But even, "I can't start today because I am having a root canal," will suffice. For other people, getting started is only a hurdle which must be crossed before successful writing can take place. These writers, many of them professionals, have solved the problem by developing personal systems for getting started. They differ from writer to writer, but they have one thing in common: they force writers to work by providing them successful ways of getting started.

In this chapter, we will look at one method for beginning a paper or report. It has worked for a lot of people, even those whose personal work-style preferences might not match well with this system. And although I can't guarantee it, it will probably work for you—if you follow it rigorously.

Figure 3 introduces a concept I call Visual Organization Systems (or VOS, for short). Taken from flow-charting techniques, these visuals will be used throughout the book to summarize procedures discussed in various chapters. The VOS depicted below outlines one approach to writing; each subsequent VOS, which deals with a particular type of document discussed in this book, would be plugged into the outline box in Figure 3.

AUDIENCE ANALYSIS

The first matter in this process over which you have some control is audience analysis. If you are lucky, this has been started by whoever assigned you the communication task in the first place. For example, you may have been told by a superior to write a report about a particular project and address it to a particular person or group. If so, continue the audience analysis along the lines discussed in Chapter 3. If this has not been the case, if you have initiated the communication, or worse, if your superior has given you no direction as to who the audience is, then audience analysis is even more necessary, and it should begin with the first step described in the previous chapter. Before you can identify what the audience knows, you have to know who the audience is.

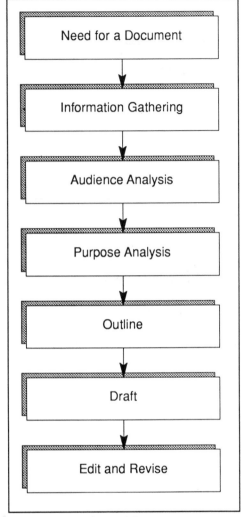

Figure 3. VOS for Writing

PURPOSE ANALYSIS

The second matter to be considered is the purpose of the project. In fact, you can think about it while you are doing audience analysis. Doing so is probably unavoidable. But since most of us cannot do two things at once, it is helpful to analyze your purpose immediately after you have analyzed your audience.

Realize that every document has a purpose. A helpful way to make yourself think about this is to write out the following sentence before you write anything in the document itself:

The purpose of this document is _____.

Then fill in the blank. Doing this should lead you to identify three important points about your planned document:

- The problem the document addresses.
- The technical issues or major points to be made.
- The rhetorical issues or what the document will do for readers.

If you cannot identify these things, *don't start writing*. You are not ready. You need to gather more information about your subject, the situation, or the audience.

After you have completed analyzing your purpose, the best way to use the results is to design the document's outline around them. If you do, outlining becomes a simple and painless task. Also, the purpose analysis can be used as an introduction to most reports. In fact, it is the best type of introduction for most reports because it orients readers to the central issues in a document and tells them specifically how the document will meet their needs.

The following example is an introduction from an analysis report. Although it is not necessary to divide an introduction into three paragraphs (each for a different aspect of purpose analysis), notice that this introduction does in fact treat each of those aspects.

Any design process begins with a clear definition of the task. In the case of a microprocessor, the task involves either controlling something or manipulating data for the means of achieving an end. The microprocessor design detailed in this report is of the second type—it is to search through 512 locations in a video game's state memory and generate the data and control signals to display the video game in its present state on an x-y vector display.

The state memory is arranged such that the data blocks for each body to be drawn are in permanently assigned locations in the lower half of the memory. The second half is set aside as a character list: sequential data blocks identify selections from a list of 38 possible characters to be displayed on the screen where specified by the coordinates in the data block.

There are two clock signals that this video display unit requires from the video game state control or "update" unit. In addition to the required access to the address lines and the data lines of the state memory, the update unit must supply: (1) a clock that signals which processor is to control the state memory and (2) a common 1.33 MHZ system clock for easier processor synchronization. The state memory control clock (V clock) must change state every 2^{14} system clocks giving each unit 12.3 ms. to perform its task. This is a necessary restriction since allowing more time results in game displays happening slower than 40 times per second, the rate at which screen flicker becomes noticeable to the eye.

In order to provide a more precise task description, I will proceed in this report with a description of the video game and how it is to appear, following with a map of state memory and the information contained therein. Following this, I will proceed to develop the system architecture designed to perform the appointed task. The result of the architecture design is the instruction set with which the program can be compiled and the specifications with which the hardware can be built. I will discuss the microprogram as a way of summarizing the architecture; then I will conclude with an outlined description of the realization and debug of the prototype.

INFORMATION GATHERING

Gathering information is the lifeblood of writing documents. Information is what readers want; delivering it is the writer's primary purpose. The introduction above points out how much can be done for readers when writers have sufficient information about audience, purpose, and subject.

Information may be gathered by a number of methods, including observation, experimentation, and literature searches. Another method for gathering information, which is not considered as often as it should be, is through fact-finding interviews. Interviewing for information is particularly appropriate for people writing about high-tech, because developments occur at such a rapid pace that the literature rarely keeps abreast of them. The best way to find information, then, is to ask for it. Chapter 5 will explore how to conduct those interviews.

OUTLINING

Outlining is something everyone should do before starting a document. Now, even though this sounds like typical academic advice, it isn't. Outlines have a purpose, beyond busy work for active junior high school students. They provide writers with a framework for the document. Doing so, they also provide the first indication whether the intended purpose is being met for the intended audience. If the the purpose is not being met, it takes far less time to rewrite an outline than it does to rewrite an entire document. For this reason alone, using outlines makes sense.

The type of outline you use is not important. This statement might be heresy to some English teachers, but it is true in the real world. An outline is only a device to help writers get organized, nothing more. Any formality, or lack of it, in an outline is strictly the business of writers; whatever they are comfortable using and whatever succeeds for them is right.

For example, outlines may be formal sentence outlines in which the topic sentences of each paragraph of every section of a report or paper are written out. This makes writing the rough draft of a document fairly easy. But there is a disadvantage to formal sentence outlines as well. Their rigidity or formality makes it less likely that editing will be as rigorous as it should be. Writers who write a draft from a formal sentence outline are reluctant to change things in the editing phase—even when things should be changed. The rough draft looks better than it is.

On the other hand, writers may opt for informal phrase or note outlines. The purpose of these outlines is to structure only the arrangement of ideas throughout the report or paper. No attempt is made to actually write the ideas out during the outline phase; this step is saved for the rough draft. The advantage to this outlining method is that the rough draft will be rough. And that's as it should be. A *rough* rough draft promotes effective editing and improves the final document. It even decreases the writer's overall work (although the opposite may seem to be the case) if we factor in the time spent rewriting a document after it has been sent out and returned because it did not work.

In addition to these common approaches to outlines, some writers (myself included) may even prefer visual or picture outlines such as mind-maps. Figure 4 is an example that I used as the basis of a 90-minute presentation I made as a consultant to a well-known computer company. As strange as an outline such as this may appear, it works particularly well for people who prefer to keep writing (or speaking) projects open-ended, people who fall into the intuitive-perceptive categories mentioned in Chapter 1. Sensing-Judging types prefer outlines that are more formal and structured. Just remember: there is no one correct way to outline. Use what works for you.

In order to help you determine what works, notice that outlines appear to be inherent in some subjects. For example, if I were to outline a physical description of the Zenith PC I am currently using, I immediately notice that I am limited by the subject. There are four physical parts to this machine: the CRT display unit, the keyboard, the CPU, and an attached printer. The only thing I have to do is to decide in what order to describe them. The order I listed in the previous sentence is a natural one—the order in which most people would notice the parts of my computer if they were standing in front of it. The display unit jumps out immediately at first glance. Next, we notice the keyboard that is below and in front of it. After that, our eyes follow a spring-cord to the CPU quietly humming away on an adjoining table. And finally, beyond the CPU, is the attached IBM ProPrinter. So at this stage, the outline might look like the one on top of page 31.

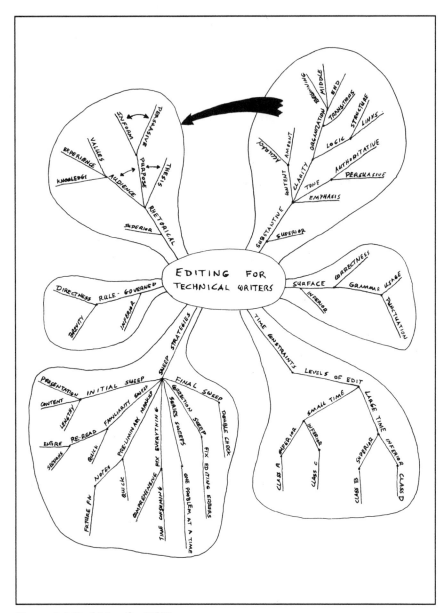

Figure 4. Mind Map for Editing

OUTLINE: ZENITH PC—PHYSICAL DESCRIPTION

1. CRT Display Unit

2. Keyboard

3. CPU

4. IBM ProPrinter.

The next thing I might do to make the outline more useful is to organize my description of each individual part. This would give me a two-level outline and a good deal more information about my subject. The CRT has a beige plastic case, a viewing area, an on/off switch, two connectors, and four controls. The keyboard has the standard arrangement of typewriter keys, a control keypad, a number keypad, an arrow keypad, and a row of function keys. The CPU has a metal case, a 5¼" floppy disk drive, a hard disk drive, and several connectors. The printer has a beige plastic case, a control panel, a tractor feed, and a continuous sheet of perforated paper. Notice in these lists, however, that I have again followed a "first-noticed" organization plan. Now my outline might look like this:

OUTLINE: ZENITH PC—PHYSICAL DESCRIPTION

1. CRT
   1.1. Beige plastic case
   1.2. Viewing area
   1.3. On/Off switch
   1.4. Two connectors
   1.5. Four controls

2. Keyboard
   2.1. Typewriter keys
   2.2. Control keypad
   2.3. Number keypad
   2.4. Arrow keypad
   2.5. Function keys

3. CPU
   3.1. Metal case
   3.2. 5¼" floppy disk drive
   3.3. Hard disk drive
   3.4. Connectors

4. IBM ProPrinter
   4.1. Beige plastic case
   4.2. Control panel
   4.3. Tractor feed
   4.4. Continuous sheet of perforated paper

If I were to go on, I would notice that several of these second-level topics could be further subdivided into third-level topics and so on.

By now, however, you should have noticed several important things about outlining. First, the order of the outline depends upon what you decide is important about the subject, and this depends upon your analysis of the audience's needs. For example, if I had chosen to do a functional description rather than a physical description, my outline would have necessarily been different. Second, if you are familiar with the Zenith PC, you know that I was outlining my description from the point of view of standing in front of each component and then moving to the back. As a writer you have to make some decisions about point of view, and that depends on your purpose. Remember: The material you include in a document and the way you arrange it depend ultimately on your purpose in meeting the reader's needs.

ROUGH DRAFT

The first draft of a document should be quick and dirty, in a word, rough. The goal is to get ideas into writing as quickly as possible, leaving editing for another time. Remember my comment about people not being able to do two things well at the same time. If you try to write and edit at the same time, you will do neither well.

But avoiding editing while writing takes tremendous willpower. If you use a computer for your drafts, that delete key is a temptation. But try to avoid it; it slows your thought processes down and forces them to be two-directional—forward and backward. Also, the result of this type of writing is not as good as it should be, either. You can print the rough draft on a laser printer, and it will come out looking great. You'll be tempted not to change a thing, and then the rough draft problems, which will naturally be there, are passed on to your readers.

A successful rough draft is messy, and the writer should be the only person capable of fully understanding it. After the rough draft is complete, leave it alone for 24–48 hours. Research has shown that this is the time required for us to forget what we wrote. After that time, we can come back to a draft and edit it from the reader's perspective, correcting problems in logic and organization. If on the other hand, we write and edit at the same time, we will still have all the ideas floating around in our heads; we will subconsciously make logic connections that are not there; and we will be too impressed by our own brilliance. In short, we will not change what we should, and the reader will suffer for it.

One final problem often encountered during the rough draft is writing roadblocks. These are dealt with specifically in a later chapter. But more than likely, if you find yourself blocked during the rough draft, you are probably not writing a rough draft. Rather, you are trying to write and edit at the same time. Don't.

EDITING

Editing, which requires locating the weaknesses in a rough draft, and the subsequent redrafting of a document make up the final stage of this approach to writing. Methods for doing this are fully discussed later in Chapter 20, but the process should be briefly mentioned here. Editing should be an organized activity. First, edit for problems of logic, organization, and clarity. Can you (or someone else) follow the development of ideas through your document? Next, edit for style. Are the sentences varied in length and structure? Are they interesting? Finally, edit for grammar and spelling. Too many poorly written documents are the result of editing done in the wrong order. Often writers get overly concerned about the nit-picking details of grammar and punctuation in a document before they have gotten the thing written clearly. Once it is written clearly, then pay attention to these details. And don't trust grammar or spell checkers either. More often than you might guess, they are wrong.

Good editing enables writers to discover areas where the purpose is not being met, areas where the development of ideas is not lucid, and at last areas where grammar, punctuation, and spelling are faulty. Don't overlook editing.

CONCLUSION

In this chapter, we have examined one process of writing from start to finish as a way of getting organized. It is not the only way, but it is a successful way. Following these steps for every document you write will come as close as anything you can do to ensuring that the documents succeed.

SUGGESTED READINGS

Rathbone, Robert. *Communicating Technical Information*. Reading, MA: Addison-Wesley, 1966.

Weiss, Edmond. *How to Write Usable User Documentation*. 2nd Edition. Phoenix: The Oryx Press, 1991.

Winfrey, R. *Technical and Business Report Information*. Ames, IA: Iowa State University Press, 1976.

CHAPTER 5

How to Get Information with Interviews

Information for reports and papers can be obtained in many ways. The most common are observation, experimentation, and reading. Observation is a valuable tool for gathering information because it is first-hand experience. The writer is present at the event that is being reported. The disadvantage with observation, however, is that it is totally dependent upon the viewpoint of the observer. Such viewpoints are rarely accurate, they are always subjective, and they vary widely from observer to observer. This is the reason that witnesses to accidents consistently give different accounts of what happened. Experimentation is also first-hand experience, but it is more accurate because its procedures are rigorously controlled. Information generated by experimentation is usually considered to be the most valid type. Design processes, and the information they generate as a by-product, are a combination of experimentation and observation. Consequently, this type of information is considered reliable, too. Finally, reading can be used to generate information. Reading is second-hand experience, though, and its accuracy is highly dependent on the accuracy of the original writer.

Another way exists to obtain useful information—interviews. When we stop to think about it, we use this technique more often than we realize. A telephone conversation that seeks to obtain information is a type of interview. So is a face-to-face conversation. The problem is that we rarely think about the interview aspect when we talk with someone else on the phone or face-to-face, and we almost never plan our conversations ahead of time.

Well-planned interviews, however, can be a valuable source of information for the person who has to write about high-tech subjects. In this chapter, we will look at one way of designing and using them.

DESIGNING AND PLANNING INTERVIEWS

Designing interviews requires us to plan them and to anticipate the reactions and answers of our interviewees. The central point to keep in mind is that we are searching for important information, and everything must be structured. As we have already seen, however, people differ in how they communicate and in how they respond to communication. As a result, audience analysis is important in designing an interview. You need to know your interviewees (or at least something about them). How receptive are they to being interviewed? Are they outwardly open or not? These issues, and others, enter into determining how successful your interviews will be.

Remember that interviews are extraverted experiences. Some people like them and others do not. You will want to take this into account when planning for and designing an information-gathering interview. For example, conducting an interview with a person who is forthcoming is easy. You basically plan a conversation to elicit information you need. If either or both the interviewer and the interviewee are introverted types, however, the process could be very difficult. With careful planning, however, the interview could still be successfully completed. In these cases, you will want to prepare open-ended questions to draw out needed information.

Preparing for an interview goes beyond scheduling a time that is convenient for you and the interviewees. The most important thing you can do is spend some time learning about the subject of the interview. If this also sounds like a variant of audience analysis, it is. Nothing will doom an interview more quickly than if the interviewees think you are wasting their time. And the way not to waste interviewees' time is to show that you have "done your homework." People are much more likely to share information with you if they are convinced that you have taken the time to learn something about their speciality. Reading is the best way to brief yourself on the subject of the interview. It is even more effective if you can read something the interviewee has written; this type of flattery never hurts.

Opening Remarks

Whether you are conducting an interview over the telephone or face-to-face, the first thing you have to do is convince the interviewee that your need for information is important and worth the time it will take to meet it. Introducing yourself and the company or division or group you represent are the bare minimums. After you have done this, it is important to get to the point of the interview immediately by stating what the problem is—why you need information. This gives your interviewee a rationale for being helpful.

Problem Background

Explaining enough background to the problem so that it orients the interviewee to your mission is important. This step requires you to explore two areas: how the problem can be specifically defined and how it was discovered. In other words, give the interviewee a clear understanding of the problem and its ramifications *before* asking for specific information. But remember: Keep this aspect of the interview short; it's still the introduction.

Interviewee Incentives

At this point, if additional convincing is necessary, it is a good idea to tell the interviewee why participating in this process is beneficial. If it is not beneficial to the interviewee, then you should rethink the situation. *It ought to be.* Rethinking usually entails only redefining the problem or re-evaluating your mission. At any rate, you must anticipate all this while planning the interview. Don't allow it to become evident during the interview.

Direct Request for Help

After the introductory remarks ask the interviewee directly for assistance. Give the person an idea of how much time it will require. If you have planned well, yes is the only possible answer—you know your subject and situation that well. Remember, however, that all this is only preliminary to the questions that will get the information you need. Don't forget to keep all introductory parts of the interview as short as possible—a couple of minutes at most.

Information Gathering Questions

To obtain the information you need, go into an interview with a list of prepared questions. The list will probably contain a combination of open-ended questions and specific questions. The open-ended questions are designed to get interviewees to elaborate on a subject, to share their range of experiences with you. The specific questions are aimed at gathering precise information.

But even with open-ended questions, make sure you have designed them to get results. Open-ended is not a synonym for unplanned. Don't say: "Can you tell me something about the new operating system?" A planned open-ended question would be more like: "I understand that the new operating system is designed to do x-y-z; could you elaborate on that?" If on the other hand, you are looking for specific information, your questions might be more like: "How does the new operating system improve read/write capabilities from hard disks?" or "How does the new operating system compare with other operating systems in its ability to read and write from hard disks?" In other words, whether your questions are open-ended or specific, lead the interviewees toward information that you can use. Prepare enough questions to cover the subject, and plan for follow-up questions, too.

Closing Remarks

At the close of an interview, let the interviewee know how important the information is (again!). Don't forget to express your appreciation for the time and cooperation. You might even add that you will inform the interviewee about the results of your communication project. And you might ask at this time whether or not a follow-up interview could be done if necessary.

CONCLUSION

In this chapter we have looked at an alternative method for gathering information for reports and papers about high technology. Interviewing is a valuable tool because high-tech industries are extremely compartmentalized. In large companies, hardware people may not know what software people are doing; programmers may not know what technicians are doing; and no one in the technical areas knows what the marketing and sales people are doing. In addition, information is produced as fast as new products are produced, and new products are produced daily. As a result, reading gives you hopelessly outdated information. Observation requires you to be all places at once. Experimentation requires you to know something about everything. That leaves only interviewing as a consistently successful method of getting useful information. Incorporate it into your bag of writing tools.

SUGGESTED READING

Stewart, Charles J. and William Cash. *Interviewing: Principles and Practices.* Dubuque, IA: William C. Brown, 1985.

CHAPTER 6

How to Explain Your Subject

Explaining oneself is the craft of clear communication. It can all be reduced to this common denominator. Rhetoric, in the classical, apolitical sense, is the body of techniques by which we explain our knowledge of a subject to an audience. Nothing about it is new; most of the ideas in this chapter have a 2400-year successful history. Nor is any of it mysterious or particularly complicated. In fact, most of it is common sense. Moreover, not all aspects of rhetoric are applicable to technical communication, but many are. Among the most useful are the following techniques: definition, classification/partition, comparison/contrast, cause and effect, deduction, and induction.

DEFINITION

There are six types of definition techniques that occur in documentation: formal/informal, operational, metaphorical, contextual, stipulative, and divisional.

Formal/Informal Definitions

Formal definitions are the most common type of definitions found in documentation. In fact, they are common to all types of communication in which the purpose is to explain something new to audiences. Correctly written, formal definitions adhere to a rigid format:

Term to be defined = class to which it belongs + differentia (information that distinguishes the term from all other members of its class)

Example: Artificial intelligence is a complex aspect of computer programming that strives to model human intelligence.

When designing a formal definition, remember that the = almost always is some form of the verb "to be," and that the differentia is the information following the word "that."

Formal definitions are most often found as the topic sentences of paragraphs, or as the lead to a new segment of information within longer sections of reports. They establish the subject and help to expand or explain it. When writers choose to expand formal definitions into product descriptions, they commonly use examples, illustrations, and further definitions. If the topic being described is complex enough, or if the treatment of it is involved enough, formal definitions may be used to begin sections, papers, reports, or even books. Look again at the first paragraph of the Introduction.

Informal definitions are actually formal definitions that have been altered for the sake of stylistic variety. Often, if you are describing a complex device, piece of equipment, or product, you would not want to begin the description of each component with the same communication technique—a formal definition. By using informal definitions in some of these instances, you can vary your sentence structure, and by doing that, you can keep readers awake. The following example shows how such short sentences can be combined effectively into a longer, informal definition.

(1) An ohmmeter is an indicating instrument that directly measures resistance in an electric circuit.

(2) An ohmmeter was used to test the circuits on the IC board.

(3) An ohmmeter, an indicating instrument that directly measures resistance in an electric circuit, was used to test the circuits on the IC board.

Notice that the third sentence embeds the formal definition of sentence one into the procedural statement of sentence two. The result is a more informative sentence, as well as one that is more stylistically interesting. Informal definitions are one way to avoid an overly simple writing style, while still maintaining logical clarity.

Operational Definitions

Operational definitions are most useful in describing processes or procedures. They do so by explaining how the process changes over a period of time or how it works, or by describing a way to measure it. For example:

A head crash occurs when a piece of dust finds its way onto a disk and disrupts the operation of the stylus, causing the system to shut down, resulting in massive erasures of information from the disk and massive headaches for computer users.

Notice that the format for an operational definition is not as rigid as the format for a formal definition. Rather than using a verb that establishes the state of something (as in a formal definition), operational definitions use action verbs. If the writer's purpose had not called for an operational definition, a formal definition would have been just as easy to use. But notice how it changes the sense of the information:

A head crash is a major disk malfunction in which . . .

Using formal definitions in procedural writing, though allowable, almost always results in wordier, weightier writing.

Metaphorical Definitions

Many people who lack imagination or the abilities of close observation might insist that metaphors have no place in technical communication, that they should be consigned for all time to the nether regions of liberal arts or creative writing. There are a few problems with such an attitude. First, all writing is creative in that a human mind is selecting words to communicate uniquely an idea to another human mind. Second, the belief that metaphors have no place in technical communication is blatantly ignorant of reality, particularly insofar as communication within the computer industry is concerned. Metaphors are everywhere. True, many of them are so overused that they have become trite, worn out, and unnoticeable, but they are there; the computer industry has an especially rich heritage of dreaming up metaphors and foisting them upon everyday communication. For example, who would think of "bug" in an entomological sense? If someone said, "I have to get the bugs out of this program," would anyone rush to call an exterminator? Of course not. Or, is a "daisy wheel" one of those flower rings most of us made during grade school recesses?

Metaphorical definitions are not methods that writers should be avidly hunting, however. To do so would be a waste of the writer's time and not very successful in the long run either. Good metaphors tend to evolve more often than they are created. And they start in everyday, on-the-job conversation. From there they become the jargon of a very limited group. If they prove to have wider applicability, their use expands to larger and larger groups until the general public uses them and they have lost most, if not all, of their original usefulness. But such is the dynamic nature of the English language; it's what makes communicating fun and challenging. Metaphors (and their understood definitions) create much of what we identify as industry jargon.

Contextual Definitions

Contextual definitions are basically truncated formal definitions. The important thing is that the definition depends on the context in which a term is being used. For example, take the word "bond." In the context of psychology, its meaning has to do with relationships. In mechanical engineering it is a fixative. In finance it is a type of investment. In chemistry it has another meaning, in physics another, and so on.

The use of contextual definitions depends on the sophistication of your audience. Writers must always be sure that readers are aware of the proper meaning of terms within the communication's context. That awareness often determines the amount of definition needed in a document, from a glossary to simple in-text definitions.

Stipulative Definitions

Stipulative definitions are a narrower type of contextual definition, used within the context of single documents or document sets. For example:

In this report, the term "chip" will always refer to a silicon wafer on which integrated circuits have been printed.

The advantage of using a stipulative definition is that the writer can preset the context and the meanings of terms when they are first encountered in a document. Used this way, an alphabetized list of stipulative definitions is a glossary. If the glossary is placed at the beginning of a report, the audience will have been informed of the specific context in which to read a document. As a result, stipulative definitions are most useful when writing to audiences that do not share the writer's knowledge.

Divisional Definitions

Divisional definitions are not really definitions at all; they are organizational techniques that can be used to expand other types of definitions, usually formal definitions. Usually, divisional definitions set up extensions of a subject, the kind often found in technical descriptions. They adhere to the following format:

X (term being defined) is composed of n parts/types.

This technique is very useful for beginning paragraphs and sections. It forces organization on the writer and expectations on the reader. The following example shows a divisional definition expanding a formal definition.

The DECmate is a stand-alone personal computer which is composed of a central processing unit, a video display terminal, and a keyboard.

If this definition were used to begin a section of a manual about the DECmate, most readers would expect the section to continue by discussing, *in order*, the listed components of the DECmate. If the writer did not abide by this forced organization, the readers would be understandably and justifiably confused. So for both writers and readers, divisional definition is a useful tool for organizing information in order to explain it.

CLASSIFICATION/PARTITION

Classification and partition are the next useful rhetorical techniques for explaining your subject. They are divisional definitions moved up one level of complexity; they adhere to strict formats, as well.

X can be classified as a type of Y.

or

X can be subdivided into the following parts: a, b, . . . n.

These two techniques are used to help organize technical descriptions of objects and processes. The following is an example of a technical description:

2. SOFTWARE

CLASSIFICATION/
PARTITION

DIVISIONAL
DEFINITION

FORMAL DEFINITION

CAUSE AND EFFECT

The basic flowchart for system software is shown in Figure 2. The software is broadly classified into two types: monitor and command processing. The monitor consists of the initializer, user in/output handler, command parser and process caller. The command processing section is a group of independent software modules which are called by the process caller to execute commands. The modularity in the software makes it easier to develop as well as add software to the system. Another command can be implemented by including an additional processing module and changing the command parser and process caller to recognize the new command.

CAUSE AND EFFECT

The monitor section was fully implemented. Since the tape system is the star of the project, the remaining software time was devoted to it. Unfortunately, problems with the read/write electronics on the tape controller board (which is not part of the system hardware) slowed down the development process. The problems were cleared up just before the deadline allowing the available software to prove that the concepts work. The tape system was not fully implemented. The other command processing modules were not implemented at all.

The following sections describe the software that was implemented. The software referred to below are listed in Appendix 2.

2.1 MONITOR SOFTWARE

DIVISIONAL
DEFINITION

The monitor software includes the initializer, user in/output, command parser, and process caller.

- PHIMON is the entry for the initializer. The stack pointer is set to top of memory (17FF Hex). Index register IX is set to checksum register address. An init message is then sent to the user. PHIMON is followed by CMDIN.

FORMAL
DEFINITION

CAUSE AND
EFFECT

- CMDIN is the command parser and process caller loop. It waits for an input character from the user, echoes it, and then checks for known commands and calls appropriate processing modules. If the character is an unknown command, an error message will be typed and the software returns to the beginning for another command. The known commands are:

  *E : Erase tape

  *F : Fast forward tape

  *I : Initiate tape

  *O : Turn off tape drive

  *P : Play tape

  *S : Stop tape.

FORMAL
DEFINITION

- SERRCV and SERSND are the user in/output routines. SERCV returns with Z flag set if nothing is received, otherwise input data is in A register. SERSND sends character in A register to the user.

2.2 TAPE SYSTEM SOFTWARE

CAUSE AND
EFFECT

Tape recordings are sequential by nature. In order to ascertain where the reading head is in relation to the rest of the tape, sector markers are recorded 1 second apart during the tape initiating. The directory of files on tape is kept in the first sector of the tape. Data files are recorded in the following sectors. Figure 3 shows a representative format of a tape with data. A file access will start the tape fully rewound, then the directory read, file location looked up, sector searched, and finally file loading.

CAUSE AND
EFFECT

Before the tape can hold any files, it has to be initiated. The formatter goes through the tape, writes sector markers 1 second apart, and then verifies the markers to make sure no bad section exits. (In later software, the bad sections will be identified and avoided when writing data.) Problems in the read/write electronics of the tape controller delayed the implementation of the tape initiating routine. Since the tape cannot be formatted successfully, file saving and reading are obviously beyond reach.

CAUSE AND
EFFECT

A major portion of software time was spent tracking down the bug, and it was corrected just before the deadline. Therefore, most of the tape system software remained unfinished. However, since the reading and writing are now reliable, the full tape system software can be implemented in two weeks.

DIVISIONAL
DEFINITION

The following routines are parts of the tape system which were implemented:

- TPSTOP : Stops the tape drive.

- TPFFWD: Fast forwards the tape.

- TPRWND : Rewinds the tape.

- TPERSE : Starts erasing the tape.

- TPREAD : Starts reading the tape. Returns byte in A register.

- TPWRTE : Starts writing the tape. Writes byte in A register.

- TPWAIT : Wastes 100 milliseconds.

- WRBLK : Writes data block to tape. Block starting address should be in HL. Block length should be in B register. Checksum is automatically appended to block.

- RDBLK : Reads data block from tape. Input buffer starting address should be in HL. Returns with block length in B register. Checksum error is signalled Z flag not set.

- TPINIT : Formats a tape. Sector markers are written 1 second apart and then verified.

The important thing to note in this example is not only how the writer made use of both classification and partition techniques to organize the information, but how this was augmented by the use of visual devices, such as bulleted lists, to make the partitioned information stand out and attract attention.

COMPARISON AND CONTRAST

Anyone who has ever been subjected to a writing course has been asked to write a comparison-contrast paper, and often the unspoken response has been "Why?" Does such a thing as a comparison-contrast report or paper exist in high-tech industries? The answer is no. But that does not mean comparison-contrast is unimportant. On the contrary, it is a vitally important technique, which appears regularly in proposals and analysis reports. For example, when you write a problem-solving proposal that analyzes several suggested methods for solving the problem, you automatically use comparison and contrast to organize your analysis. This technique may not exist as a pure form for a report or paper, but it is required to organize your thoughts for certain types of reports or papers.

As is the case with other writing techniques, there is a limited number of ways that you can structure comparison-contrast. It can be organized as a point-by-point analysis (some people refer to this as alternating), or it can be organized as a subject-by-subject analysis (called block by some). The best way to understand the difference is to view the two methods diagrammatically.

Imagine that you want to compare two things (A and B) and that you want to make three statements about each thing. If you arrange your material as follows (A1, B1, A2, B2, A3, B3), first a statement about thing A, then a statement about thing B and so on, you have designed an alternating comparison. On the other hand if you arrange your material (A1, A2, A3, B1, B2, B3), all statements about A followed by all statements about B, you have designed a block comparison. But remember that with block comparisons you have to provide clear transition between the blocks so readers can tell how you got from one place to another.

The important question is when do you use which type of comparison. Block comparison works for all types of comparisons, long or short, simple or complex—as long as you take care of the transition between blocks. Alternating comparison should be limited to short, simple topics—usually no more than two or three things compared and no more than two or three statements about each one. This is a very important criterion. For example, imagine that you are going to compare 3 things (A, B, and C) and that you will make seven statements about each, *and* that you are going to use alternating comparison. Off you and the readers go—A1, B1, C1; A2, B2, C2; A3, B3, C3; A4, B4, C4. . . . At about this point readers start to notice something funny. They begin to see the structure of your information while failing to pay attention to the information itself. No matter how skilled a writing artisan you become, you never want this to happen. Craft never supercedes information in technical writing.

CAUSE AND EFFECT

The rhetorical technique of causal analysis, or determining the causes and effects of events, is the fundamental organizational technique of analytical reports. These reports may analyze problems, or they may explain research and development activities at the conclusion of a project. But at the core of these reports is the treatment of cause and effect.

Causal analysis can be accomplished in any of the following formats.

Single Cause >>>>>>>>>>>>>> Single Effect

Example: I dropped the chalk. It broke.

Single Cause >>>>>>>>>>>>> Multiple Effects

Example: I dropped the chalk. It broke and scattered pieces on the floor.

Multiple Causes >>>>>>>>>> Single Effect

Example: I tripped over the chair and dropped the chalk. The chalk broke.

Multiple Causes >>>>>>>>>>> Multiple Effects

Example: I tripped over the chair and dropped the chalk. The chalk broke and scattered pieces on the floor.

The philosopher in you might argue that all events cannot be reduced beyond the fourth format, multiple causes leading to multiple effects. In reality, this is correct. Like the young child listening to the Genesis story, we can always ask, "What happened before that?" But remember: Don't ever confuse technical communication with reality. The two are not the same, no matter how objective you strive to be. Technical communication is a re-creation of reality based on the needs of an identified audience. A large amount of detail selection goes on in the writing of a report or paper, and our purpose is to cull the irrelevant details so that we can explain our subject. Depending on those purposes and our audience, we might not feel a need to analyze a person's breakfast in order to determine the causes of a software programming error. Nor, perhaps, would we need to report every single event that led to the failure of a lab test, just those that had a significant bearing on the result.

Be careful, however, because this brings us to an important aspect of cause and effect: determining what is important and what is not. Although there is a large body of information called formal logic that deals with this subject, we do not have to be concerned with it in detail. Suffice it to say that after we have determined our communication purpose, after we have analyzed our audience, and after we have gathered information, we have to determine what causes and what effects are vital to the needs of our readers. That is then the causal analysis to communicate.

DEDUCTION / INDUCTION

Deduction and induction are two organizing techniques that are mirror images of each other. Deduction requires that the writer state a general principle and then give examples to support it. This chapter is a good example of deductive organization. I began with a statement defining clear communication, and I have spent the rest of my time explaining how it can be achieved by examining different examples. Induction is just the opposite. The writer strings together a series of examples or incidents, each edging the reader a little further along. Then comes the punch line, or the statement that ties all the incidents together. Notice that I used the terminology of joke-telling; most jokes are organized inductively. Consider the following story:

> I grew up in a small town in rural North Carolina where on work days people farmed and on off days they either fished or sat around telling lies about farming and fishing. A man we all knew as "Old Pops" was a local fishing legend. While most of us would come back after a day on the lake with four or five fish, Old Pops never failed to return with less than thirty or forty. Bass, catfish, perch, anything that swam Old Pops could catch. And he did, regularly. When the rest of us told lies about our fishing, Old Pops just sat silent. After all, his actions spoke much louder than our tales put together. Eventually, though, his talents attracted the attention of the game warden. And one day the warden dropped by to suggest that he and Old Pops go fishing together just to see how Old Pops managed his success. What could Old Pops say? A week later, he found himself in his boat on the lake with the game warden. As the warden was baiting up, Old Pops pulled a stick of dynamite from inside his shirt, lit it, and tossed it over the side. Seconds later, there was a muffled "whumpf," followed by about a dozen fish floating motionless to the surface of the lake. The game warden, flabbergasted, almost speechless with surprise and indignation, cried, "Old Pops, you can't do that. Don't you know it's against the law, and I'll have to arrest you for it?" Old Pops said nothing. He just reached inside his shirt, pulled out another stick of dynamite, lit it, and tossed it to the game warden's end of the boat. Then he said, "Warden, are we going to sit here and talk all day, or are we going to fish?"

What makes this particularly good induction is that the principle is never explicitly stated, even at the end. Rather, it is implied. As a technique for explaining information, however, induction does not have as much potential for use in writing about high technology as does deduction. It is also much harder to control and retain the reader's interest using induction. But it can be used. The following is a more serious example.

> On October 3, at 4:15 a.m., a widespread power outage occurred in the east part of the city, where our primary research and development facility is located. This resulted in massive losses of data from our PDP-11 computer. The specified method for dealing with these sorts of unforeseen disasters is to have multiple back-up copies of data stored on tape or disk. Following the power outage, however, we learned that back-ups had not been made in approximately two weeks. Consequently, two weeks worth of work was lost on Project Antelope. Aside from the fact that this will put the project further behind schedule and cost estimates, the lesson which R & D personnel must learn from this is that back-ups must be made daily.

Notice that in this example inductive organization is combined with multiple causes >>>> multiple effects organization. For most cases, the use of inductive order in writing about high technology should be limited to descriptions that are short and introductory in nature.

CONCLUSION

In this chapter, we have looked at a variety of commonly used rhetorical techniques. You already use these techniques in your everyday conversation, but it is important to be aware of their planned use in writing so that you control them and not the other way around. The time to examine whether you are using these techniques effectively, however, is not when you are writing the first draft of a document. Do it while editing. If you think too much about the technique while you are writing a first draft, you will not be concentrating enough on the information to be communicated. Remember: form follows function. Or more explicitly: technique never takes the place of content; it enhances it.

SUGGESTED READINGS

Billings, Millard F. "Prose Pollution." *IEEE Transactions on Professional Communication.* Vol. PC-16, No. 1 (1973), pp. 11-14.

Bolz, Roger W. "Communicating the Significance of Technology." *Automation.* Vol. 14 (July, 1967), pp. 55-61.

Brandt, E. N. "In Favor of Four-Letter Words." *Chemical Technology.* (Sept., 1971), pp. 528-31.

Colby, John B. "Paragraphing in Technical Writing." *Technical Communication.* Vol. 18, No. 2 (1971), pp. 13-16.

CHAPTER 7

How to Use Graphics with Reports and Papers

Too often writers overlook the importance of including graphics in their reports and papers. Correctly done, graphics are attention getting and informative. They can carry much more information per space in a document than the same amount of text can. And if there is one definite trend in writing about high-tech subjects, it is an increasing reliance on visual communication, which aids all readers, but is particularly useful in international situations because less translation is necessary.

There are two kinds of graphic communication—tables and figures. They differ in format and in the way they are incorporated into the text of a report or paper. For example, formal tables are numbered and have a title; all of this information is placed *above* the table. Figures also have numbers and titles (sometimes called captions), but the information is placed *below* the figure. The only difficulty is remembering what is a figure and what is a table. Actually, it's simple: everything that is not a table is a figure. Furthermore, there are only two types of tables—formal and informal. And only one of those, the formal table, has numbers and titles. In this chapter, the most common forms of graphics found in writing about high technology will be examined and explained. Each has its best use; writers must determine the nature of the information that is to be communicated and choose the appropriate graphic accordingly.

INFORMAL TABLES

Informal tables are simply lists. As such they are not numbered in a sequence throughout the report or paper. Rather, they are an extension of the text. But remember, they should be physically separated from the text they accompany by a sufficient amount of white space before and after the list and by additional margin space to the left and to the right of the list. This physical separation calls attention to the information in the list, and it is what makes an informal table a visual technique.

This book has a number of informal tables in it. Figure 5 is an example. Remember that the figure number and title are for the graphic in *this* text. The original did not have either.

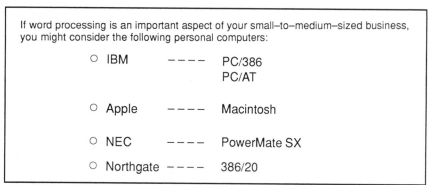

Figure 5. Informal Table

FORMAL TABLES

Formal tables require more formatting than lists. They are usually separated from the text by top and bottom rules (lines) and by a table number and table title, both of which are placed *above* the table. Formal tables should have clear column and line headings. They should accurately describe the types of information in the table. If necessary, both column and line headings may have subheadings. The individual cells for data, or information, should be spaced adequately so they are not crowded and hard to read. If an explanatory note is needed for any information contained in the table, it is placed immediately below the table as a footnote. If the entire table was copied from another source, the title of the table receives a reference—either with a superscripted arabic numeral or a numeral in parentheses. Figure 6 is an example of a formal table.

TABLE 2 (BINARY CODED DECIMAL)				
DECIMAL DIGIT	BINARY 8421	BINARY 2421	EXCESS CODE	GREY CODE
0	0000	0000	0011	0000
1	0001	0001	0100	0001
2	0010	0010	0101	0011
3	0011	0011	0110	0010
4	0100	0100	0111	0110
5	0101	0101	1000	0111
6	0110	0110	1001	0101
7	0111	0111	1010	0100
8	1000	1110	1011	1100
9	1001	1111	1100	1101

Figure 6. Formal Table

LINE GRAPHS

Line graphs are used to show changes in the state of something over a period of time. In order for them to communicate effectively, the graph axes must be labeled clearly and descriptively, with units of measure also marked. In multiple line graphs, the lines should be differentiated with symbols (which are explained in a key) rather than with various dotted and dashed lines. This will enable you to use the graph more effectively as a slide in a presentation, should that be required. In all graphs, plot lines from graph paper must be removed before inclusion in the final document. You can do this by either tracing the graph in ink or by working it out in black ink on blue-line graph paper and photocopying it. The blue grid lines will not copy. An exception to this rule would be the inclusion of working graphs in an appendix. There the lower quality of the visual is not as important.

Figure 7 is an example of a multiple line graph.

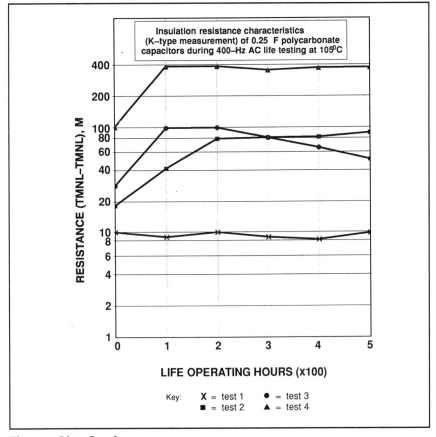

Figure 7. Line Graph

BAR GRAPHS

Bar graphs are used to show size relationships of items in a comparison. Bar graphs may be designed on axes similar to line graphs, but neither axis will signify a change in either state or time. Once again, the axes must be clearly labeled. Any units of measurement are also marked, and the actual measurement is usually listed at the top of each bar.

Figure 8 is an example of a bar graph. (Please note: Information in figure is fictitious and is used solely for illustration.)

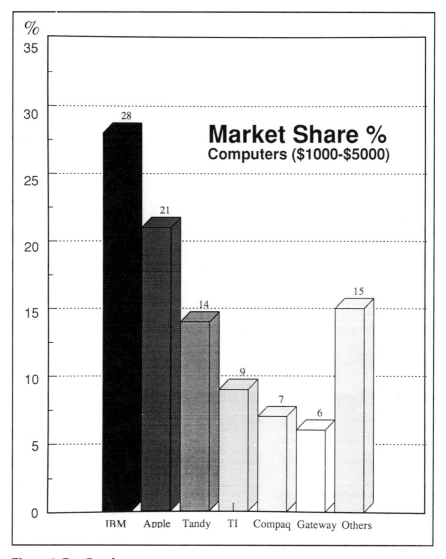

Figure 8. Bar Graph

DIVIDED BAR GRAPHS

Divided bar graphs are similar in appearance to bar graphs, but divided bar graphs compare percentages of a whole rather than relative size. Bar labels and percentages are often placed within the bar when space allows it. If there is not enough space, this information is placed outside the bars, and faint lines are drawn to connect the information to the appropriate bar.

Figure 9 is an example of a divided bar graph. (Please note: The information in this figure is fictitious and used solely for illustration.)

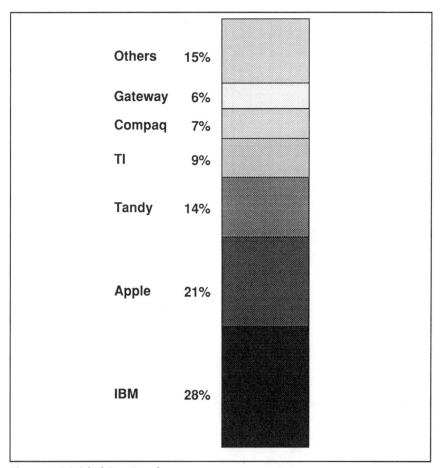

Figure 9. Divided Bar Graph

CIRCLE GRAPHS (PIE CHARTS)

A circle graph, or pie chart, serves the same purpose as the divided bar graph: it compares percentages of a whole. It is probably more effective visually than a divided bar graph for this purpose because readers can get a clearer sense of the whole. Circle graphs are complete; bar graphs look as if they could be extended. With the circle graph, labels and percentages can be placed inside each wedge if space allows. Often it does not, and even when it does, labelling is made difficult by the odd angles of the wedges. For these reasons, labels and percentages are usually placed outside the circle graph and connected to the appropriate wedge with a line. Remember, also, that the largest wedge begins at 12 o'clock and moves clockwise.

Figure 10 is an example of a circle graph. (Please note: Information in this figure is fictitious and used solely for illustration.)

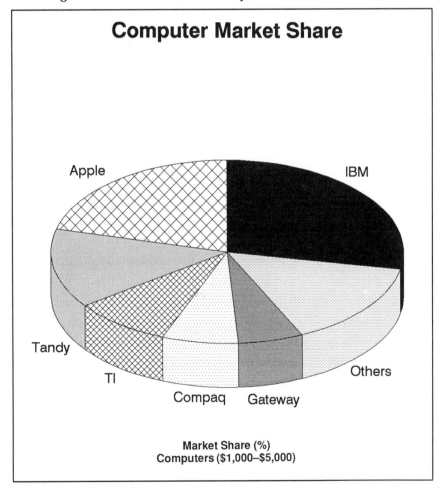

Figure 10. Circle Graph

PICTOGRAPHS

Each graphic we have examined so far can be drawn and used by anyone who has access to a compass and a ruler. Pictographs are within the domain of graphic artists. They are essentially bar graphs composed of pictures rather than bars, and they are very effective visually.

SCHEMATICS

Schematics are particularly important to writers of reports and papers in the high-tech industries. They are used to show the relationship of parts, a type of visual outline. For schematics to work, however, they must be very clear and absolutely uncrowded. Labels and headings should be placed where they can be read. Circuit drawings and flowcharts are familiar schematics.

Figure 11 is an example.

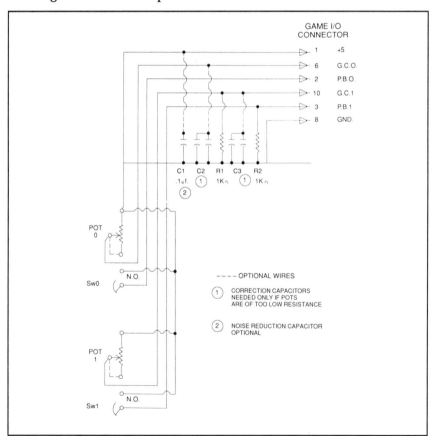

Figure 11. Schematic

ILLUSTRATIONS

Illustrations are the final type of graphic to be considered. They are two- or three-dimensional renderings of an object. To do these successfully, you should know something about drafting. Otherwise, leave these up to the company graphics person or a free-lance artist. If you do them yourself, be sure to orient the illustration in a way that makes sense to the reader. Also, label all important parts clearly and give appropriate dimensions. It is important that readers, particularly unsophisticated readers, are able to tell whether what is being illustrated is the size of a nuclear reactor containment vessel or a silicon wafer.

Figure 12 is an example of an illustration.

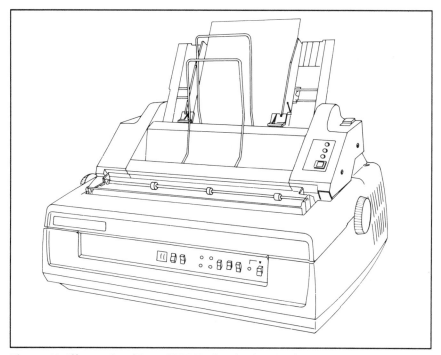

Figure 12. Illustration (From NEC Technologies, Inc.)

PLACEMENT OF GRAPHICS

One point, which has not been mentioned so far, is the placement of graphics. This is a controversial issue. For reports, if the graphic helps explain an important point to the readers, place it as close to that point as possible. Usually this means on the same page. If the graphic is included for reasons of documentation accuracy (to cover yourself), it is acceptable to place it at the end of the report in an appendix. For articles and papers submitted to magazines and journals, note the position of the graphic (by name and number) in the manuscript. Then place the graphic at the end of the manuscript. The editors will see that it is properly placed.

One final note: Since the first edition of this book was published, dozens of graphics packages for personal computers have been developed. Some are superb; many are garbage. Make sure that if you are planning to use a graphics package that the graphics produced still adhere to the conventions of accurate, usable graphics.

CONCLUSION

In this chapter, a variety of graphics that writers can use with reports and papers have been examined. When chosen and rendered accurately, with your purpose and audience in mind, they can be vital additions to written communication. Be sure that every graphic you use is clearly explained in the text. This chapter ends with a figure that will help you determine which graphic is most appropriate.

GRAPHICS USE CHECKLIST	
Rhetorical Purpose	*Type of Graphic*
Describe Organization	Pictograph, Flowchart
Compare/Contrast	Pictograph, Pie Chart, Bar Graph
Describe Parts	Schematic
Classify	Table, List, Pictograph
Describe Change of State	Line Graph, Bar Graph
Relate Date to Constants	Line Graph
Describe Process	Pictograph, Flowchart
Describe Steps in a Decision	Flowchart, Pictograph
Describe Multiple Responsibilities	Flowchart
Describe Proportions	Pie Chart, Bar Graph
Describe Relationships	Table, Line Graph
Describe Causation	Flowchart, Pictograph
Describe Entire Object	Schematic, Illustration, Photograph
Present Raw Data	Table, List

Figure 13. Graphic Selection Chart

SUGGESTED READINGS

Angel, Edward. *Computer Graphics and Graphics Programming.* Reading, MA: Addison-Wesley, 1989.

Fetter, W. *Computer Graphics in Communication.* New York: McGraw-Hill, 1965.

Foley, James D., et al. *Computer Graphics: Principles and Practice.* 2nd edition. Reading, MA: Addison-Wesley, 1990.

Garland, Ken. *Graphics Handbook.* New York: Van Nostrand Reinhold, 1966.

Gerken, J. Ellen. *The Brightest Computer-Generated Design and Illustration.* North Light Books, 1990.

Grieman, April. *Hybrid Imagery: The Fusion of Technology and Graphic Design.* Watson-Guptill, 1990.

Haskins, George. "The Graphics Designer—A Concept," *21st ITCC Proceedings.* Washington, DC: Society for Technical Communication (1974), pp. 115-18.

Schmidt, Calvin F. and E. Stanton. *A Handbook of Graphic Presentation.* New York: Wiley-Interscience, 1979.

CHAPTER 8

How to Use
Writing Tools

The explosion of technology available to everyone who writes makes this chapter a necessary addition to this book. In the past seven years, more tools have been developed than any one person could ever hope to use, and in fact, some of those tools do things that none of us should ever need to do. If there is a catchphrase to this chapter, it is "Beware of Glitz!" In other words, when you are deciding on the writing tools you need, look at what is available and buy only that which meets your needs. This chapter will help you decide what works and what doesn't.

WORD PROCESSING

Word processing is where the writing technological boom began. Microcomputers and the software for them have become affordable, or at least available, for almost everyone. These tools allow writers to outline, draft, polish, and revise their work without ever committing print to paper. Then, when everything is right, a laser printer can make our words look more professional than was ever possible before. But "look" is a key word here. Lightning quick technology does not absolve any of us from the responsibility of paying attention to the details described elsewhere in this book. New technology may tempt us to do just that.

I am convinced that although revision of text is much simpler using word processing, many people (because of the beauty of laser-printed text) simply don't do it anymore. The demands of writing about high technology require that information be generated quickly, and word processing allows that. But we should not forget the principle of getting our communication correct before we commit it to the readers.

Many people find it helpful to write a draft very quickly, print it out on a less expensive dot-matrix printer, edit it, and then revise it. That works for me, and a lot of professional writers agree. Try to avoid the temptation to write and edit at the same time. As I mentioned earlier, most of us are not good at those sorts of mental gymnastics.

You'll notice that I am not going into detail on what type of word-processing software you should use. There is probably no quicker way to start an unresolvable controversy than to get into that. Some people prefer Wordstar, some Microsoft Word, others Xywrite, and others MultiMate. All word processors have benefits and limitations. You should look them over, try them out, and see which you will learn to use most easily and which meets the needs of the kind of writing you do. The trend is toward WYSIWYG software (pronounced "Wizzywig"), or What You See Is What You Get. The theory is that the more the screen resembles the paper, the easier the word processor is to use and understand.

One final word about word processing software and microcomputers. Don't trust them! I have been writing professionally for over 16 years and this past year that point was brought home—painfully. I had finished writing two new chapters for the second edition of this book just before Christmas. One of those chapters is titled "How to Use Writing Tools." In my haste to get out of the office and home for Christmas vacation, I neglected to do something that I almost automatically do—make a backup copy of what I had just finished writing. You guessed it: I returned in early January to discover that my computer had had a CPU failure. All the work was lost. Now as I am rewriting this chapter from notes and memory, you can believe I won't take that kind of shortcut again. Don't let it happen to you. Save important documents on a hard disk and back them up on floppy disks. Make a printed copy and keep it, too. Having been burned once, I now make copies on separate floppy disks and store one at home and the other at the office. Don't be lulled into thinking it could not happen to you. It could.

DESKTOP PUBLISHING

Desktop publishing is one step above simple word processing. With this tool, all sorts of power to manage text is at the writer's fingertips. Primarily used for its visual capabilities in producing such things as newsletters and manuals, desktop publishing allows for relatively easy full-page design. You can choose from a myriad of fonts, type sizes, and layouts. The only problem is that this often puts more variety at the writer's command than is necessary, or even than makes sense. Fourteen different fonts on a page might be a *tour de force* of desktop publishing, but it is lousy communication. Xerox's Ventura Publisher and Aldus PageMaker are industry standards for desktop publishing. Both are powerful and complex, and can produce fine-looking documents. Remember, however, that desktop publishing is still only a tool, although a mighty powerful one. Use it to make your information more accessible to readers, not to dazzle them with the output of your technology. There is still no substitute for clear, concise writing.

HYPERTEXT

What if the majority of your writing tends to involve the design of online documentation? If that is the case, then hypertext is your tool. Hypertext, an electronic means of cross-referencing information, provides readers/users with easy access to large volumes of information in an online environment. Basically, it is a database that lets you connect screens with associative links. Hypermedia refers to the ability to link text, graphics, sound, video, and computer programs. These tools are best used when you have readers who will need to "browse" through your information, following one topic to other related topics. In other words, it is useful for readers who might not know the names of what they are seeking. The associative structure also promotes learning by giving readers control over the sequence and content of the information they are seeking, and by extension, may even promote greater retention and motivation.

There are several benefits to hypertext. First, the users see only information that is relevant to them. In traditional, paper-based documentation readers have to wade through a lot of print searching for the material they need. In traditional online documentation, they do the same thing by scrolling through screens. Second, one hypertext "document" can serve many different audiences and audience needs. Different levels of sophistication and different reading paths are available to users. Third, hypertext can cut the costs of doing documentation by making the information modular. This eliminates the needless repetition in various sections of a paper document.

If you are interested in writing in a hypertext environment, there are several things you should consider. First, make sure that what you are using is truly hypertext and not a glorified global search. A true hypertext system provides associational links between information, links that will function consistently each time the reader uses them. A glorified global search simply treats a hypertext link as a highlighted word and then locates another use of that word in the database; as a result, you might not wind up in the same place twice. If the reader modifies the document (another hypertext option), the likelihood of having a global search model work is even less.

Second, you might want to consider the usefulness of being able to link text, graphics, audio, video, and traditional databases. The use of different media will depend on the subject matter you will be committing to the hypertext environment, but as I have pointed out elsewhere in this book, the trend in technical communication is toward a greater integration of these various types of media. Be prepared.

Third, make sure that any hypertext system allows readers to find their way around. Can readers go backward as well as forward in their browsing? Can they see a trail of where they have been? Is there a way to show the hierarchy of the document's structure? The point of hypertext is to manage information. Are readers well oriented to where they are and where they can go? Are indexes available? What about other methods of retrieving information?

Fourth, is the system easy for you, the writer, to use? Can you create links easily? Can you integrate text and graphics effectively?

Fifth, does the hypertext system integrate well with other applications on your computer system? Yes, even in the last decade of the twentieth century, it is still necessary to ask this question. You want to make sure you are buying a tool, not a dinosaur.

Finally, the bottom line—price. Hypertext systems are new products, and you will find competitive pricing. Just be sure you know what you are getting, that it meets your needs and the needs of your audience.

SPELL CHECKERS AND GRAMMAR CHECKERS

These tools are touted as the end to most writers' problems. Don't believe it. Now I know that some of you are committed to these things, naively believing that since you never received good grades in English composition or, especially, spelling, that these tools are life savers. They're handy, all right, but not infallible. Don't forget to check the spell checker. The reason you have to do this is that they fail in certain situations. Every spell checker has a limited vocabulary or dictionary. While some of these are immense, they are not infinite. Real words that are not in the software's dictionary will be flagged as misspelled by the spell checker. Closely related to this problem is the fact that misspellings will not be flagged if the misspelling matches a word that is *in* the dictionary (for example, *effect* and *affect*). Another real possibility is that the dictionary in your spell checker might be wrong, that it might contain misspelled words which no one knows how to spell (like "occuring," which actually has two r's). Using spell checkers is like doing a rough draft. Always read over the copy for errors.

Grammar checkers are even more unreliable. For one thing, grammar is considerably more complex than spelling. As a result, it is more difficult to design a software program that does a thorough job of checking a writer's grammar. Another problem is that there is less widespread agreement about correct grammar. Turn a grammar checker loose on a famous author's work and watch it locate error after error. Acceptable standards for grammatical usage change more readily than spelling standards. For example, is it "None are . . ." or "None is . . ."? What about *who* and *whom*? Before you buy a grammar checker, you should think about buying a good grammar handbook. They're cheaper . . . and more reliable.

CONCLUSION

In this chapter, we have taken a quick look at the tools of the writing trade. They are continually improving, making the task of sorting out what's what difficult for you. One way to keep abreast of the material is to read the journal *Technical Communication* regularly, because it contains analyses of new products. Another is to visit any reputable software dealer. But keep in mind what you need, so you avoid being sold glitz.

SUGGESTED READINGS

Barrett, Edward, ed. *The Society of Text: Hypertext, Hypermedia, and the Social Construction of Information.* Cambridge, MA: MIT Press, 1989.

——————. *Text, Context, and Hypertext.* Cambridge, MA: MIT Press, 1988.

Cole, Bernard C. *Beyond Word Processing.* New York: McGraw-Hill, 1985.

Johassen, David H., ed. *The Technology of Text: Principles for Structuring, Designing, and Displaying Text.* Englewood Cliffs, NJ: Educational Technology, 1982.

Stibic, V. *Tools of the Mind: Techniques and Methods for Intellectual Work.* Amsterdam: North Holland, 1982.

Thorell, L. G. and W. J. Smith. *Using Computer Color Effectively: An Illustrated Reference.* Englewood Cliffs, NJ: Prentice-Hall, 1990.

Zinsser, William. *Writing with a Word Processor.* New York: Harper and Row, 1983.

PART 3

How to Write a Paper or Report

CHAPTER 9

How to Organize
a Paper

High-tech industry professionals often have the opportunity or the desire
to publish outside their own particular organizations. There are many
reasons for doing this: fame, money (provided the journal pays for the
contributions), professional advancement, or because someone made you
do it. Whatever the reason for submitting an article to a trade journal or a
paper to a professional journal, the challenges (and the rewards) are
categorically different from those encountered in writing internal reports.

The first problem that confronts the author is defining just what is
being written. Is it a paper, article, or report? While there is some confusion
over the differences among these three types of writing, the typical
distinction between papers and articles is that a technical paper is formal,
intended for a relatively limited audience and published in a professional
journal (such as one of the IEEE publications), while an article is informal,
intended for a wide audience and published in trade journals (such as
Electronics Purchasing). Although papers and articles might communicate
the same information as reports, reports are generally classified into
various types of formal documents: proposals, feasibility studies, interim
reports, and so on. There is, however, enough overlap in the forms of
communication that the following chapters on writing papers introduce
topics common to reports very well.

AUDIENCE

First, there is the problem of audience. It is large, anonymous, and
very difficult to analyze. Trying to write for such an audience is akin to
writing in a closet with the door closed. You don't know what's on the
outside. Even if you are comfortable with the audience analysis system
presented in this book, you quickly realize that it does not provide as much
information in this environment as it does for internal reports. Answers to
the audience-analysis questions are automatically vague. You can't iden-

tify readers by name and position, and even if you obtained a copy of the circulation list for a journal (not a bad idea), it would only do you a little good.

The audience issue for articles and papers is bleak, but not hopeless. The best way to learn about your prospective audience is to *contact the journal editor and ask.* Each journal is intentionally aimed at a specific market; in fact, with the high-tech field being what it is, there are many journals for each market. Let the editors do some of your audience analysis for you. In addition, be sure to ask for a sample copy of the journal so that you can use it to analyze the audience.

ORGANIZATION AND STYLE

Once you have decided who the audience is, analyzed papers or articles that have been published in your target journal, and determined what information would interest your readers, you have to solve the problems of organization and style. You will have to fit your article or paper into the mode expected by readers of the journal. It's the only way. You won't be able to change editorial policy, so to be published in your target journal you will have to play the game by their rules. Contact journal editors to ask for their submission requirements and style guides.

Once again, examine successfully published papers and articles. Look at such things as sentence structure, sentence length, use of personal pronouns, the passive voice, illustrations, photographs, charts, tables, the use of headings within the paper, anything that might affect the way you present your information. Remember also the two ways of perceiving— intuition and sensing. Use headings and orientations liberally throughout your text so intuitive readers can skim. Provide plenty of details within paragraphs and sections so that sensing readers can closely follow your train of thought. Also, find out about the accepted referencing styles. The editors can answer all these questions, allowing you to spend your time more profitably by gathering information and organizing it. The more you know about what the editors expect of a submitted paper or article, the more likely you will be to get an acceptance.

WRITING THE ARTICLE OR PAPER

Based on what you find out by examining the journal, you will be faced with a different sort of communication problem with different solutions. Writing a paper or article about high technology is not the same as writing a report, even if the topic is the same. There are different skills and tasks required, all of which must be mastered by you to be a success-fully published writer. The next three chapters are devoted to the most difficult tasks of writing articles and papers—writing the discussion, the conclusion (or exit), and the lead. You might wonder why they are in this order. Why not start at the beginning? In fact, you are. For a writer, the beginning is the middle. You cannot clearly lead readers to an understanding of your subject if you do not first know where you are going yourself. Beginning in the middle is time and task effective. It allows you to write as soon as you have your information, without wandering off on tangents and into blind alleys. After you have written the middle and the end, the hardest part—writing the beginning—will be much easier.

You might also wonder why the discussion focuses mainly on writing articles. One reason is that the intended readers of this book are more likely to write articles than papers. Another and more important reason is that the format for articles is much more flexible, more open to the design whims of individual authors. Those of you who are faced with writing a paper will find formats rigidly prescribed by the target journals. A quick examination of the journal will also show how the individual sections of a paper are to be written.

Figure 14 depicts a general approach for organizing articles about high technology intended for trade journals.

One final thing, rejections. Expect them. They are a reality in the lives of all professional writers. They are not affronts to your integrity, nor to your abilities. Most often they simply mean that what you have to say does not fit with the editor's idea of who the journal's audience is. Look for another journal and try again. If you are willing to make the effort, even write a follow-up to the rejection, asking for more specific information as to why it was rejected, and asking whether or not you may resubmit it, provided necessary changes are made. This does not work all the time, or even perhaps most of the time, but if it works once it is worth the effort.

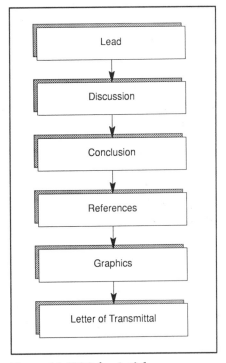

Figure 14. VOS for Articles

CHAPTER 10

How to Write the Discussion or Body of an Article

Before we can consider how to write the discussion or body of an article, we should consider what makes this part of an article work. Once again, the answer is organization. The body of an article is what supplies readers with the specific information they might be seeking. It is the place for facts and figures, nuts and bolts, or chips and gigabytes. Just as it will be important to tell readers what you are going to do before you do it, it is equally important to do it after you have told them you would. This is how a well-designed lead and body of an article should work together. The body must present the information forecasted in the lead, and it must present it so that readers can understand and use the information. Otherwise, you are wasting a busy reader's time. That is why it is safer to write the body first; then you know what to include in the lead.

The basis of all well-written discussions is order. Three common orders will be considered in this chapter: (1) reader-centered order, (2) writer-centered order, and (3) subject-centered order. Notice that each of these orders emphasizes one of the three parts of all communications—audience, writer, and message.

READER-CENTERED ORDER

A reader-centered discussion or body of an article seeks, as its main goal, to satisfy the needs of the reader. But, you say (especially after having read the chapter on audience analysis), isn't that the goal of all communication? Yes, it is; but the reader-centered discussion puts *more* emphasis on reader needs than do the other discussion orders. The following outline depicts a reader-centered order for the body of an article.

Outline of a Reader-Centered Discussion

1. The first section of a reader-centered discussion presents specific information about the topic of the article. It defines the problem or what is

at issue in the article and provides a brief statement of its resolution. It can then suggest to the reader what is to be expected in the rest of the article if this has not been done already in the lead. If the article and the lead are quite long, the first section of the discussion might repeat (briefly) a forecast of the remaining structure of the article, even though it was initially mentioned in the lead. As we saw in the section covering purpose statements in Chapter 4, the lead plus the first section of a paper's discussion constitute a purpose statement introduction.

2. The second section of a reader-centered discussion presents criteria, or assumptions, that will be used to evaluate the validity of the writer's conclusions. This is a logical extension of the first section's resolution of the central issue of the article. Establishing these criteria erects a frame around your work, limiting what your readers can take issue with. Such limits also make your job of organizing the information and the readers' jobs of understanding it easier.

3. The third section of a reader-centered discussion (usually this contains several subsections) provides the support for your conclusions. There are two possible organizations for this information: from most important point to least important and from least important point to most important. Each way has advantages and disadvantages. From most important to least important is easier for the readers to understand. You have told them what they most need to know early in the article, tying it in closely with the lead and the first part of the discussion. But if you do this, there is a chance that the readers' interest will lag toward the end of the discussion as your material gets less and less important. From least important point to most important is more dramatic and suspenseful; the readers' interest tends to build throughout the article. But most articles in the high-tech industries do not have drama and suspense as two of their primary aims. In addition, this arrangement makes it difficult for readers to follow the writer's train of thought. It is easy for readers to become confused. So, what is the answer? In most cases, the best arrangement for articles about technical subjects is from most important point to least important. Just remember that you should not belabor your points toward the end of an article to the extent of boring your readers. As your material becomes less interesting, your writing pace (and the reader's reading pace) should quicken. Be sure to look at Chapters 19 and 20 to see how to do this.

4. The fourth and final section of a reader-centered discussion restates the most important point the writer is trying to make, perhaps showing

how the point has been supported in the article. This section should lead quickly and smoothly into the exit of the article.

The following example presents excerpts from a reader-oriented discussion. The entire discussion was too long to reproduce here, but the four parts of it are nonetheless identifiable in these excerpts.

PROBLEM DEFINITION

The mealing condition evident on Printed Wiring Assemblies can be directly related to the card and system exposure of a humidity test, as Printed Wire Assemblies in systems not exposed to humidity do not exhibit mealing. Additionally, all the required contractual testing has shown that the functional performance of GTE products is not affected by mealing on assemblies. With the functional performance of the product unaffected and the mealing being relatively discrete, even though it covers a large area of the assemblies, the consensus was that GTE does have mealing but it is of a minor nature and that the long-term reliability of the assemblies will be unchanged.

The plans of the task team were therefore directed down two paths: first, to verify the integrity of the process to clean flux (ionic contaminents) that could affect reliability from the assemblies; and second, to perform investigations and testing to determine what is causing the mealing.

CRITERIA

Specification requirements for mealing are found in MIL-P-28809 which is called out by MIL-P-11268 and MIL-STD-454.

Mealing is identified as a major defect in MIL-P-28809. However, as will be shown later, it is called major under the assumption that ionic contamination is the cause.

SUPPORT

The most important investigation was determining the level of ionic contamination by verifying the effectiveness and accuracy of the Omega meter.

Specification MIL-P-28809 requires a resistivity reading of 2.0 Megaohms/cm or greater. Testing samples per MIL-P-28809, Paragraph 4.8.3., and in the Omega meter, average readings of 11.0 Megaohms/cm and 10.1 Megaohms/cm respectively were obtained. This is a negligible difference and well above the specification requirement.

Other testing conducted up to this date has shown mealing can be created on bare boards that were cleaned with processes other than aqueous and on boards that were not exposed to flux and wave solder. This result supports the assumption that assemblies are being properly cleaned and directs testing away from the solder/cleaning process.

SUMMARY OF MAIN POINTS

From investigations to date, the government has established that the cleaning processes are effectively removing ionic contamination from the Printed Wire Assemblies. In addition, GTE's experience from in-house and field testing has exhibited no reliability or functional performance degradation due to the presence of discrete Printed Wire Assemblies.

WRITER-CENTERED ORDER

A writer-centered discussion or body of an article seeks to portray the process the writer or writers followed during the research and/or development of the article. For the most part, this order is restricted to subjects that are procedural in nature, and can be structured as a narrative. When the subject lends itself to this order, or the writer's purpose suggests this order, it can be a very effective way of organizing the body of an article. It draws upon the direct experience of the writer with the subject. For an excellent example of this structure, look at Tracy Kidder's *Soul of a New Machine*, a popular, nonfiction account of introducing a product at Data General. The following outline shows how writers might organize information in a writer-centered discussion.

Outline of a Writer-Centered Discussion

1. The first section of a writer-centered discussion presents how the writer discovered the topic. Depending upon the topic, this might be a section on the need for a new software product or a section on the writer's day-to-day activities or whatever. This section extends the lead and begins the narrative.

2. The second section of a writer-centered discussion develops the narrative by describing the writer's experiences with the topic. This section is usually the longest part of this type of article. For it to be successful, it generally follows a chronological order and leads to the final section.

3. The final section of a writer-centered discussion presents the culmination of the narrative, usually giving the readers some insight and understanding, a resolution of sorts, of what the writer has been through. This section should lead smoothly to the exit of the article.

Writer-centered articles are generally entertaining as well as informative. As such, we are more likely to find them in the popular press than in trade journals. A consistent (and excellent) exception to that is the trade journal *Design News*. You will find writer-centered narratives from engineers and designers in that journal on a regular basis. One last point about writer-centered articles: the market for them is highly competitive and the success rate for publishing them is fairly low. But they are enjoyable to write, a break from the type of writing most of us find ourselves doing on a regular basis.

SUBJECT-CENTERED ORDER

This order is the one most commonly used for the bodies of articles or for the discussion sections of reports. It is serviceable and easy to use for any subject. Because it is used so frequently, this order will not make the subject stand out. Writers, however, can generate emphasis through a variety of other stylistic devices, which are considered in a later chapter.

While this may sound as if I am downplaying the importance of this order, it should not. You will find yourself using this order more than the other two combined. It is that flexible, and it works that well. Another advantage of the subject-centered order is that it can be combined with the writer-centered order. If your topic is broken down into subtopics and one or more of those subtopics is procedural in nature, you might find such a combination advantageous to your article. The following outline depicts the general-to-specific organization of subject-centered order.

Outline of a Subject-Centered Discussion

1. The first section of a subject-centered discussion presents the first subsection of the topic. The topic should have been clearly subdivided in the lead so that the readers can anticipate the order of the discussion, and the discussion should treat the subtopics in the order they were listed. A common mistake is for the writer to neglect to do this, resulting in the complete befuddlement of the readers. In this type of discussion structure, too, the writer can arrange subtopics from most important point to least important and vice versa.

2. The second section of a subject-centered discussion, and every section thereafter, presents the remaining subtopics in the order in which they were listed in the lead. The final section of this type of discussion should lead the reader to the conclusion, which is almost always a summary.

The following selection is an example of a subject-centered discussion. Notice how the discussion is simply a catalog, almost a list, of the information.

SUB-TOPIC 1

In the initialization mode, the LISP Machine processor issues commands to the frame buffer circuit board. The two commands initiate frame storage or display. The remaining commands are issued whenever the LISP Machine is cold-booted.

SUB-TOPIC 2

In the frame storage mode, the board takes analog input from a television camera or videotape recorder, digitizes it, and stores it in memory. The user indicates to the LISP Machine which type of device is supplying input so that videotape sub-mode may be entered if necessary. This sub-mode causes the board to synchronize itself with synchronization pulses generated by the videotape recorder, whereas in regular mode the board generates these pulses and supplies them to the television camera. When the board receives the command to store a frame, it waits for the top of the frame to appear, as indicated by the sync pulses, stores the frame, and enters display mode. No processor accesses to memory are allowed while the frame is being stored.

SUB-TOPIC 3

In the display mode the board accesses each location in memory. . . .

CONCLUSION

In this chapter, we have looked at common ways used to organize the bodies of articles. Remember that even though your audience might share with you considerable expertise on your subject, flawlessly clear organization will always lessen the readers' work and enhance their appreciation of you.

CHAPTER 11

How to Write the Exit

Writing a good exit, or conclusion, is the second hardest task in writing an article. (See Chapter 12 for the hardest task.) And unfortunately, there are no formulas that guarantee successful exits all the time. But good exits are vital to your communication, so time must be spent developing techniques that work for you. If it is true anywhere in the field of writing that practice is the only road to success, it is true in the writing of exits. The more you practice, the more you will develop what has been called a "felt sense" as to when they are right. In other words, you will just know intuitively when an exit is right, when it works for you and for the readers. This may sound like typical writing-teacher hocus-pocus, but it isn't. Trust me.

Or if you don't regularly trust what you are told about writing, even though formulas for successful exits don't exist, it isn't as bleak as that may make it seem. Exits tend to follow an order of their own, and we can categorize them as follows in this chapter.

SUMMARIES

Summaries are the most common type of exits written for articles in the high-tech fields. A summary conclusion of a reader-centered article might reintroduce the central issue of the article and restate the solution more fully than it was initially forecast in the opening section of the discussion. In a writer-centered article, it might restate the objectives of the writer's work or the problems or needs that caused the writer to do the work in the first place. In a subject-centered article, it might simply restate the main points made in the paper.

One potential problem with using summary exits is the tendency to confuse them with a part of a report that is also called a summary. They are not the same. A report summary is a distinct section intended for a distinct audience. If you think of it as an executive summary and realize that it is placed at the beginning of a report, there should be no confusion. The purpose statement introduction discussed in Chapter 4 can easily be expanded into an executive summary.

Another source of confusion, even if we limit our discussion to articles and papers, is that many people call an abstract a summary and vice versa. An abstract also differs from a summary exit; it is a synopsis of an entire report or paper, placed before the beginning of the paper and intended to be used independently of the paper. For example, imagine that you are doing some literature research that requires examining many papers. The abstract/summary, if well written, will enable you to determine whether or not you have to read the entire paper because it will contain the important aspects of the paper.

The following is an example of a summary exit to a writer-centered report. It is an important example because it shows that writing about computer technology often occurs outside high-tech industry in places where we might not expect it, in this case an elementary school.

CONCLUSION

Because of the influence of the computer revolution, our school board has directed us to build, equip, and manage a computer learning center at Pineville School. The school board further directed us to develop a computer teaching curriculum. In order to accomplish this goal, I was instructed to write a proposal detailing the steps needed for such a project.

I began by studying all the unused rooms in the building, their location, size, and general condition. The custodian and I inventoried and inspected every piece of surplus furniture in storage. I have spent considerable time with Radio Shack salespeople determining the following: what hardware is available that is compatible with what we own, what price Radio Shack will charge and what extended services are available. Random House publishers allowed me to inspect and sample many of their educational programs.

Based on my study, I conclude that it is reasonable and practical for us to build a computer learning center at a cost that is within our funding limits.

Using three summer workshops and inservice programs, we can do the following: train inexperienced teachers in BASIC, develop a curriculum to teach the children BASIC, buy software to supplement our reading and math programs, and establish a committee to develop long-term goals.

LOGICAL CONCLUSIONS

In a persuasive or argumentative article, the writer will most likely want to lead readers to a logical conclusion. At the end of the article, the readers will have reached the same conclusion around which the writer organized the topic. Although each type of discussion organization can be used for this type of conclusion, it is most commonly found with reader-centered discussions.

APPLICATION OF BASIC POINTS

The exit to an article might directly show the readers how they can use the information presented in their own companies, their own jobs. This, too, is most often found with reader-centered discussions, but it will work equally well with subject-centered discussions.

FORECAST OF FUTURE EVENTS

In high-tech trade journals, this is probably the most important and most desired type of exit. It provides readers, particularly management and design engineers, with information that they most need to know— where the industry is going in the next few months or years. This type of exit can be applied to each of the discussion orders we considered. It works very well with reader-centered discussions. It works well with subject-centered discussions. And it works particularly well with writer-centered narratives.

CONCLUSION

In this chapter, we have examined how to write good exits to articles. Everything that you have read can be transferred easily to writing conclusions for reports and memos. Remember, however, that the tendency is to overwrite the conclusion, to write beyond the stopping point. Only through practice and good editing will you develop a feel for knowing when to quit.

The following is an example of an exit from an article about computer technology. Notice that it is a forecast of future events.

> Now that its computer network is on line, the Medibank Private Fund is embarking on a new $1 million+ media campaign that attempts to persuade Australians that "all health insurance funds are not the same." The campaign, which includes more than 130 television spots and insertions in Australia's three top magazines, will have ads tailored, for the first time, to young singles as well as to families.
>
> Capturing the healthier, and more profitable, young singles market will be crucial as Australia's health insurance and medical industries continue to change. The elimination a year ago of free hospitalization in Australia intensified competition among funds to be sure, but with medical costs having risen by 140% in the past year alone, Medibank Private and its competitors are finding the marketplace more challenging than ever.

CHAPTER 12

How to Write the Lead

As I suggested at the beginning of Chapter 11, getting started is the most difficult task in writing. Part of the problem is simply writer's block, that tendency for all of us to freeze up at the thought of putting words to paper. That problem will be dealt with in a later chapter. But another important part of the difficulty in writing leads is not knowing what an effective lead is, what it does, and how we can construct it. Writing leads last enables you to avoid many of these difficulties.

Effective leads bring the topic of a paper to the reader. In technical communication, they orient readers to the topic so that nothing which follows the lead is a surprise to the readers—at least not in terms of the subtopics covered. Now, perhaps we should consider exactly what I mean by this. Although your topic might be revolutionary or surprising on its own accord, you should not wander down tangents in an article. In other words, you should not include information in the body that was not forecasted by the lead. That's what I mean by no surprises.

Just as there were a number of conclusions that can be used for articles, there are a variety of possible leads. We will consider these in the rest of this chapter.

THE OFFER OF SOMETHING NEW

This lead is used in many trade journal articles as an effective way to generate interest among readers. Almost all readers look to trade journals for the newest wrinkle in their specific field. If that is what you are presenting in an article, then you want to emphasize that fact in the lead. This lead works well with reader-centered discussions and with subject-centered discussions. It also ties in well with exits that forecast future events.

> America's competitive edge, dulled in recent decades by imports and outdated business tactics, is being honed anew. In both design studios and boardrooms, quality has taken on a new role—from that of lip service in advertising campaigns to new prominence as the ultimate business strategy.

Article continues by describing the new strategy
Dana L. Gardner
Design News, Volume 46, Number 3, 2-12-90, page 110.

SUMMARY OF PAST DEVELOPMENTS

This lead (yes, it's yet another type of summary) serves as a way to orient readers to a topic with which they might not be completely familiar. It is an attempt to show the main topic of an article as the natural outgrowth of what has occurred previously in the industry. Although it can be used with all three discussion organizations, it is a particularly effective way to start a subject-centered article.

As the Voyager spacecraft head out of our solar system and set their sights on interstellar space, Bill Layman can take special pride. His design work on the space probes more than 12 years ago played a major role in a successful journey that yielded new scientific data and breathtaking photographs of the planets.

Article goes on to describe the design work.
Carlton F. Vogt, Jr.
Design News, Volume 46, Number 3, 2-12-90, page 90.

COMPARISON OF THE OLD WITH THE NEW

This is a good lead for a paper that is introducing a new development to the industry. What better way to generate reader interest than to describe your topic by relating how it is better than what it is intended to replace. This way of subtly combining informative writing with advertising is something we all know is done but something that is not talked about in public very much—at least in the public of technical writing. This type of lead also lends itself very well to use with subject-centered discussions, although it is usable with the other two as well. Tracy Kidder uses it effectively to begin *The Soul of a New Machine.* The following article also uses this type of lead:

"EUREKA! EUREKA!" Archimedes is said to have shrieked after he stepped into his bath and watched the water rise. In that instant, Archimedes realized that a body immersed in fluid is buoyed up by a force equal to the weight of the displaced fluid. Legend has it that he ran through the streets naked he was so excited about discovering this principle.

Surprisingly, not much has changed in the intervening 2000 years. Water still rises when you get in the tub, and engineers still get some of their best ideas when they least expect it. The creative mind works in mysterious ways.

Article continues by examining how creativity can be harnessed best in work environments.
Tristram Korten
Design News, Volume 46, Number 3, 2-12-90, page 138.

CONCLUSION

In this short chapter, we have examined possible leads and what they can accomplish. This section is not a magic hat, however, from which writers may always pull out the right rabbit. The choice of a lead is closely tied to the writer's purpose, to the article's formality, and to its subject. Only careful analysis of these factors will provide the best basis on which to construct a lead.

Remember that the techniques discussed here apply to articles. A formal technical paper almost always begins with a direct approach to the subject and the audience. The purpose statement introduction, which was discussed in Chapter 4, is an excellent way to do this.

The following is a sample lead from an article about computer technology. Notice how the offer of something new is used to organize this lead effectively, without calling attention to the structure of the lead itself. This should be the writer's goal: an unobtrusive but always present structure in every communication.

> Even to people who do not program computers themselves, the value of computers as tools for many scientific, engineering, and educational applications is apparent. Until recently, though, the time spent learning how to use the computer, writing programs, or training a programmer often seemed to outweigh the benefits of computer use.
>
> Typically, someone with a series of calculations to do needs a set of specific, specially written programs to have a computer perform the calculations. Someone in this situation might adapt his needs to something "do-able" with commercially available programs. If that is impossible, he must either learn how to write the necessary programs himself or hire and explain his needs to a computer programmer.
>
> All these options have long lead times before a program can be used effectively. Also, unless the person has previous programming experience and writes the series himself, he will probably have to compromise his standards and use something that will not do exactly what he wants.
>
> TK!Solver is a personal computer product allowing effective computer use without a programming background. Developed by the creators of the Visicalc spread sheet program, TK!Solver is a "custom programmer," flexible enough for a variety of applications. It is a single program written to solve a wide range of problems instead of a collection of programs each for a single problem. This makes it very adaptable to the user's needs. Already designed for flexibility, TK!Solver can be tailored to the individual without fundamental alterations.

Notice how this lead draws the readers into the topic, whets their appetites for more information, and *leads* them to the type of information they might expect (and should find) from the paper: namely, how the program can be used by people without programmer backgrounds and how it is flexible to meet a variety of computing situations.

SUGGESTED READINGS FOR CHAPTERS 9, 10, 11, AND 12

Benrey, Ronald M. "How to Get Your Technical Article Published." *20th ITCC Proceedings*. Washington, DC: Society for Technical Communication (1973), pp. 143-44.

Day, Robert A. *How to Write and Publish a Scientific Paper*. 3rd edition. Phoenix, AZ: Oryx Press, 1988.

Dodds, Robert H. *Writing for Technical and Business Magazines*. New York: Wiley-Interscience, 1969.

Mitchell, John H. *Writing for Technical and Professional Journals*. New York: Wiley-Interscience, 1968.

Sachs, Harley L. *How to Write the Technical Article and Get It Published*. Washington, DC: Society for Technical Communication, 1976.

PART 4

Additional Types of Writing

CHAPTER 13

How to Write Memos

In terms of how frequently memos are used in the computer industry, or in any technological business, they are the most important type of documentation you are likely to write. They are the everyday workhorses for transferring information from one person to another. They may vary from extremely informal handwritten notes to formal, typed interoffice memorandums (IOMs), they may even include electronic mail and on-screen messages. Although most writers of memos do not consider the possibility, memos even have an afterlife as the source material technical writers use to compile manuals. Clearly, they are vital to the efficiency of a company's communication. Figure 15 shows how a writer's approach to designing a memo might be organized.

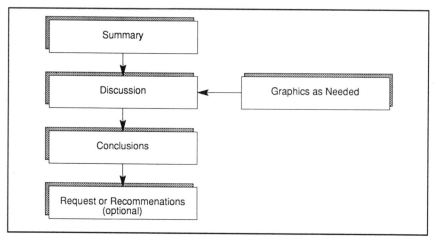

Figure 15. VOS for Memos

TYPES OF MEMOS

When we consider the number of memos written by computer professionals during their careers, it seems wise to simplify the discussion of techniques used to organize and write memos. One way of doing this is to classify memos according to the purpose they serve.

(1) Recommendation Memos (either requested or self-initiated)

(2) Progress Reports

(3) Informative Memos

(4) Information-Requesting Memos

The design of each type of memo varies slightly (in the case of progress reports, greatly), depending on these classifications. The rest of this chapter will be devoted to discussing these memo designs.

Recommendation Memos

A recommendation memo is used when writers wish to make a recommendation to the reader concerning some problem or issue which the writer has investigated. It might help to think of recommendation memos as informal proposals. As was suggested above, the information contained in these memos can be either requested by the reader or initiated by the writer. If the potential reader has requested an investigation that will lead to conclusions and recommendations, the memo generally adheres to the following design.

Introduction (reader-centered)—The introduction summarizes the problem as the writer understands it and provides a brief statement of the recommended solution.

Discussion—The discussion or body of the memo presents an evaluation of all the alternatives the writer considered, including the pros and cons of each.

Conclusions and Recommendations—This section recommends a solution, analyzes it, and discusses important aspects of implementing it.

If the writer initiates a recommendation memo, feeling that the intended audience has a need for the information presented, it generally adheres to the following design.

Introduction—A three-part purpose statement introduction that supplies background orientation to the issue being reported on, a definition of the specific problem at hand, and a brief statement of the recommended solution.

Discussion—An evaluation of all the alternatives the writer considered, including pros and cons of each.

Conclusions and Recommendations—A recommendation of a solution, a full analysis of it, and the presentation of important aspects of implementing it.

Progress Reports

Progress reports are issued at predetermined intervals during an ongoing project. These intervals vary depending on the needs of the company. They may be daily, weekly, monthly, or quarterly. Regardless of the intervals, progress reports are used for one purpose: to report the progress made on a project during the interval covered by the report. They can do this most concisely when they adhere to the following design.

Introduction—An orientation to the status of the project by describing its background, including what has been accomplished prior to the present.
Discussion—An analysis of the accomplishments and problems during the interval being reported on.
Future Work—A statement of the work planned for the next interval.

Informative Memos

These memos are issued at the discretion of the writer to provide readers with information they "need to know." They are usually short, to the point, and often informal. Even so, for them to work, they should be well organized. The following design provides that organization.

Introduction—A three-part purpose statement that orients the reader to the subject of the memo, provides a brief description of the contents of the memo, and suggests the action the reader should take with regard to the information presented in the memo.
Discussion—The presentation of the information.
Requests—An *optional* section used if the writer wishes information from the reader pertaining to the contents of the memo.

Information-Requesting Memos

These memos are issued when the writer has identified a need for specific information. Usually, they too are short. The following design makes it likely that the request will be noticed easily by the reader.

Introduction—An orientation to the writer's needs, including a reason why the memo was written and why it was sent to the reader.
Discussion—A direct request for the information needed.

CONCLUSION

Just like this chapter, memos should be as concise and direct as possible. That does not, however, stipulate an optimum length for memos. Some formal IOMs can run as long as 10-20 pages, including attachments.

If you write a memo that long, however, look for opportunities to subdivide long sections with subject-oriented subheadings. A good rule of thumb to follow is never to confront the reader with a complete page of unbroken text. Use headings and graphics to give readers a rest.

Also, remember that in most cases you will know the readers of your memos personally. That will allow you to take into account their personal preferences for receiving information. By using the designs presented in this chapter, combined with your knowledge of the audience at your company, you can ensure clear communication in your memos.

SUGGESTED READINGS

Beene, Lynn and Peter White, eds. *Solving Problems in Technical Writing.* New York: Oxford University Press, 1988.

Burnett, Rebecca. *Technical Communication.* 2nd edition. Belmont, CA: Wadsworth Publishing Company, 1990.

Dumont, Raymond A. and John M. Lannon. *Business Communications.* 3rd edition. Glenview, IL: Scott, Foresman, 1990.

CHAPTER 14

How to Write Specifications

For everyone involved in the design phase of the computer industry—hardware engineers, software engineers, technicians, and programmers—specifications are the most important documents to be read or written. They dictate design.

Obviously, the audience for specs is highly technical, but that does not mean that the audience necessarily shares the knowledge of the writer. This is particularly true when specifications cross major professional lines within the computer industry; for example, a specification written by marketing people for hardware engineers, or even a spec written by hardware engineers for software engineers, may not be useful if the writer assumes too much of the reader. So, even though the language of specifications is precise and the treatment of the subject is thorough, great care must be taken to ensure that the specs can be used by the intended audience. Nothing wastes more time within an organization than having to figure out poorly written specs. The situation is even worse when work has to be undone or redone because of bad specs.

TYPES OF SPECIFICATIONS

For our purposes here, specs can be subdivided into four types (there are other sub-types, but that is a more specific treatment than we need in this book; for additional information, look at the list of suggested readings at the end of this chapter):

- requirement specs
- functional specs
- design specs
- test specs

Although the names for these specifications may differ from company to company, each type can be found in computer companies that design, produce, and market a computer product. All four types will be discussed in this chapter.

Requirements Specifications

The requirements specification should be the first step in the design phase for a new product or for a product that is being updated or changed. Often, people in the marketing division of a company are responsible for determining what the market desires. At the lowest common denominator, this is what market research is about.

The result of market research is the requirements specification. In it, the marketing people attempt to specify what the market is looking for, what people or companies who use computers would find useful and would like to have. A former student of mine who now works for Lotus Development Corporation, and who writes these types of specifications, refers to it as "writing corporate fictions." Requirements specifications specify something that does not exist, and because they are often written by people who are not design engineers, they are likely to be the most general of specification types. They can do no more than provide marketing's best effort to describe what would be a profitable product.

Even so, they are extremely valuable. They provide the design group with a place to start. Often, they will contain enough information so that readers can see relationships to past technology. In order to do that, requirements specifications should contain the following as a minimum:

- Product Definition—As accurate a description as can be written by marketing about the desired product. It should answer the question: "What is it?"
- Functions List—A description of what the desired product should be capable of doing. It leads to the next type of specification.
- Cost—A ballpark estimate of what the desired product should cost to be competitive in the marketplace.

Functional Specifications

In most organizations, after the requirements specifications have been written, a group is formed to study the desired project. This study is usually broken down into hardware functions and software functions, and it leads to the writing of both hardware and software functional specifications. These specs will form the basis for the highly precise design specifications.

Hardware functional specifications as a rule contain the following:

- Functional Description of the Product—A precise description of the purpose, use, and operation of the product. It might reference related documents; discuss user, performance, and compatibility requirements; and present enhancements or options.

- Configuration Specification—Specifies such things as how the product's components are to be interfaced with each other and with other available products.
- Electrical Description—The specification describes the electronics that will be used to accomplish the product's capabilities.
- Physical Characteristics—This contains a precise description of each of the product's components.
- Standards—Specifies how the desired product should fit into existing company standards.
- Environmental Requirements—A description of how the product will be used and under what conditions.
- Diagnostic Requirements—A description of the testing and evaluation requirements for the product.
- Power Requirements—Describes what sort of power source will be required to operate the product.
- Cost Target—Establishes what the product should cost the consumer.
- Maintenance Cost Target—Establishes what the expected maintenance costs are likely to be. Usually the costs per month are figured.
- Resource Requirements—Specifies what resources will be needed to design the product.
- Documentation—Outlines the necessary documentation for the product. Depending on the product and its uses, this could be many things. Usually, it describes the manuals that will accompany the product.
- Risks—A discussion of the risks inherent in pursuing the design, development, and production of the product.
- Assumptions—Describes the underlying assumptions that can be made about the product and the process of designing, developing, producing, and marketing it.
- Unresolved Issues—Even in documents this thorough, issues can remain unresolved. These issues should be presented and discussed.
- Glossary—Ensures that readers from a variety of technical and nontechnical backgrounds can understand and use the specifiction.

Software functional specifications are similar, but there are enough differences to warrant examining them. Sections which repeat sections of the hardware functional specification will not be explained again. Software functional specs usually contain the following:

- Functional Description of the Product
- Product Features—This section describes the capabilities of the product in detail.
- Environment
- Dependencies—This is an elaboration of what the implementation and use of the software will depend upon.
- Physical Characteristics
- Risks

- Assumptions
- Cost Target
- Maintenance Cost Target
- Resource Requirements
- Documentation
- Glossary

The level of detail increases dramatically between the requirements specification and the functional specifications. That detail will become even more specific in the design specifications.

Design Specifications

Design specifications are based on functional specifications. The goal of design specs is to provide a detailed design of each of a product's features. These specifications are best begun before the design process starts and updated while the design process continues. Design specifications are later used as a basis for test plans and user documentation.

Hardware design specifications generally contain some version of the following components:

- Introduction—A three-part purpose statement introduction that explains the need for the product, lists the specific features of the product, and forecasts the contents, organization, and use of the design specification.
- Applicable Documents—A list of the documents that contain information pertinent to the product. Such a list is absolutely essential to the technical writers who will be producing the documentation manuals for the product.
- Functional Description—A detailed description of the product's functions, what it is designed to do, and how it is designed to do it. Block and circuit diagrams are used extensively in this section, but make sure that they are not a substitute for clearly written text.
- External Interfaces—Specifies all interfaces which apply to the product.
- Detailed Design—Details the design of individual aspects of the product's functions. It is the most detailed section of the design specification.
- Programming Considerations—Describes all aspects of the hardware with which a programmer would come into contact.
- Power Requirements—Describes the assumed power requirements for the product.
- Reliability—Explains how reliable the product is designed to be and what is to be expected of it with regard to service and maintenance.

- Diagnostic Considerations—Describes the testing and evaluation requirements for the product.
- Standards—Explains how the product will fit into existing company standards.
- Environmental Requirements—Describes the operating conditions that are assumed for the product.
- Glossary—Defines potentially unfamiliar terms for a wide range of readers who might have to use the specifications.

The software design specification is also similar to the hardware design specification. But as was the case with functional specifications, here too there are differences that need to be explained. The software design specification is also used as a basis for testing and as the source for user documentation. To meet these two purposes, software design specifications should contain the following:

- Introduction
- Applicable Documents
- Functional Description—This section should be subdivided into however many functional features the software has.
- General Design —The design section details the way in which the software design objectives are met. It is the most detailed section of the software design specification, including such material as data structures, data flow, program relationships, and so on.
- Memory Requirements, Performance, and Restrictions—Details how the software fits into and uses computer memory. It assesses the performance of the software and presents any restrictions that might apply.
- Product Requirements—Discusses such matters as security, usability, and installation and maintenance requirements.
- Test Strategy—The test section presents any helpful suggestions that could be used in developing a test plan for the software.
- Deviations from Functional Specifications—Sometimes, changes are necessary. This section describes them.
- Interface
- Glossary

User documentation for computer products is universally lambasted for being unreadable and unusable. Much of the problem lies in specifications, which later become source material for technical writers. If computer professionals would take the time to write and appreciate the importance of thorough design specifications, anticipate one of the eventual audiences of these specs (technical writers and the users), and follow a version of what has been presented thus far in this chapter, user documentation would improve drastically.

Test Specifications

Before a product can be marketed, it must be tested to see how well it will perform under market conditions. This procedure also should be specified consistently across a company. In order to do that, test specifications should contain at least the following:

- Introduction—A three-part purpose statement introduction helps to describe the purpose of the tests and to forecast the contents of the specification.
- Applicable Documents—These documents might describe test procedures on similar products that have been designed and developed in the past.
- Description of the Unit to Be Tested—Identifies and describes, thoroughly, the test unit.
- Testing Method—Provides a step-by-step description of the testing procedure.
- Precautions—Details any special care that must be taken in the testing phase. The same degrees of precaution apply here as they do for procedures. See Chapter 15.
- Glossary—Here the glossary defines potentially unfamiliar terms for the people who will be conducting the tests.

CONCLUSION

In this chapter, we have examined four broad categories of specifications common to the computer industry. Minor differences will occur from company to company and from industry to industry, but there is one universal: a high degree of accurate, exhaustive description. This is both necessary and desirable. Too often specifications are hastily written with no thought for their eventual use. Writers see only the present and figure that the limited group of readers can work through turgid specs. This is untrue, and it forces unnecessary work upon all readers, leading to projects that fall behind schedule and run over budget, as well as to documents that no one can use. When more and more companies are pointing out that the information products (documents!) can affect the marketing success of a company's computer products, clear specifications are vital.

SUGGESTED READINGS

Crown, David F. "Ten Commandments of Writing Readable Specs." *Specification Engineering.* Vol. 35, No. 3 (1976), pp. 54-57.
Eberlin, Fred E. "What to Tell Your Engineers about Electronic Procurement Specifications." *Technical Communication.* Vol. 22, No. 1 (1975), pp. 2-4.

Freier, Martin. "Effective Specification Writing." *Technical Communication.* Vol. 22, No. 1 (1976), pp. 14-16.

Sides, Charles H. "Computer Documentation." In *Technical and Business Communication: Bibliographic Essays for Teachers and Corporate Trainers,* Charles H. Sides, ed. Urbana, IL: NCTE, 1989, pp. 329-40.

CHAPTER 15

How to Write Procedures

About 12 years ago, suffering from the delusion that I was "good with my hands," I bought my wife a bureau in a box. Like most people when they start work on a kit, the first thing I did was open one end of the box and dump all the contents out onto the floor of our apartment. The last thing that fluttered out was the instructions. The first thing they said was, "Remove contents from box in an orderly fashion." The wisdom of this was immediately apparent. There was no parts list, no handy little plastic bags containing the hardware, just a lot of wood and metal scattered on my floor. The next thing I noticed was that the instructions had only three steps. There were at least forty objects on my floor, and even though the project was simple (one that could be done in "3 hours," or so the box had said), I had my doubts. So, I did what most people do when they find themselves in this predicament: I threw the instructions away. Three days later, something resembling a bureau had grown in our apartment, but there was a small problem. The last piece of wood, which looked as if it ought to be a restraining piece fitted across the back of the bureau, was 4 inches longer than the top of the bureau. Nothing suggested it should fit diagonally, a solution which would have looked odd anyway, and I did not want the extra four inches sticking out into the room to grab people as they walked by. So again, I allied myself with kit builders in this kind of predicament: I got my saw, sawed the excess off, and nailed the piece in place. The bureau stood up and did bureau things for years.

For several years, that was the worst set of instructions I had ever encountered. But a couple of years ago, I found the all-time winner. I had contracted for an addition to be built onto my house. The contractor said I could save $4000 by adding a woodstove instead of the fireplace my wife and I wanted. This was a persuasive argument, so I went out and bought a woodstove I could install myself. I knew there was going to be some problem when the woodstove dealer brought the product to my Jeep using a forklift. I told him that was nice, as he placed the woodstove in my car, but that I did not happen to have a forklift at home. No problem, he responded; simply take the stove apart in the box, carry each piece into the house, and reassemble it there. Sounded easy to me, so I left for home with a brand-new, shiny, efficient Norwegian stove that cost me just under $1000. I did what the dealer said, and after I had scattered the woodstove around my new room, I located the instruction manual at the bottom of the box. The first five pages were in Norwegian. The next four were in Swedish, the next four in German, and the next three in French. The last page contained the English translation, which consisted of just two sentences: Step 1. Install woodstove, and Step 2. Have fire inspector check installation. Certainly, one could not fault the writer's brevity.

These stories do serve a purpose in this chapter. They identify the single most common problem of poor procedure writing: procedure writers who have no idea who they are writing for and under what conditions the unfortunate user will be doing the procedure. Consequently, the worst technical writing tends to be "how-to" writing. I have even heard people say that they are convinced the persons who write procedures have no idea how to perform them. While I would not go that far, such an attitude does suggest a public relations problem for those of us who write procedures. Our information product often does not inspire confidence that even we know what we are doing. That is reason enough for considering how to write procedures. This chapter, then, will do just that: examine a method for planning, writing, and testing procedures.

PLANNING FOR PROCEDURES

Before readers can be told how to do something, they must be told what the something is and what its purpose is. In other words, give readers a rationale for doing the procedure. Doing so suggests to readers that you have considered them and what they are up against in their day-to-day activities, that you don't want readers wasting time trying to figure out what you want them to do. This consideration has an obvious positive effect: readers are more likely to follow the procedure accurately.

You can further enhance their sense that you considered them by taking into account their different needs for information. Nowhere is the difference between personality-based preferences for information more important than in the writing of procedures. Sensing readers prefer a detailed, step-by-step chronology while intuitive readers need only overviews and a general sense of what they are going to do. Be sure to provide readers with both types of information.

Procedures, regardless of the format, may do any or all of the following:

- give steps for operating something
- give steps for assembling something
- give steps for trouble-shooting, repairing, adjusting, or maintaining something
- give steps for unpacking or shipping
- give steps for ordering parts, optional attachments, and so on
- furnish a parts list
- teach skills, as in a training manual

The important fact that you must realize when planning and writing procedures is that you become a *teacher*. You are the expert, and it is your responsibility to impart knowledge to readers who know less about the subject than you do. If it is true of any type of writing in the computer industry, it is especially true of writing procedures: the written product has value only insofar as it can be used.

One of the ways you can ensure the usability of your procedures and a way you can instill confidence in your readers that *you* know what you are talking about is through the language you use. Use the *command* voice, rather than the descriptive voice. An example of that is the preceding sentence; sentences that begin with verbs are in the command voice. Notice that the descriptive voice (for example: "you should use the command voice") suggests that there might be alternatives to what you want done, that the reader only "should" use it, not that he or she absolutely has to use it, no questions asked. The command voice is the voice of authority. As long as you don't fail your readers with inaccuracies, the command voice is convincing.

Vocabulary is also important in procedures. The procedures must be written in the terminology of your readers. Impressing them with your education or vocabulary is worthless. If readers are to follow a procedure accurately, they at least have to be able to understand the individual steps. This is just one more indication of the importance of audience analysis and adaptation. As a writer of procedures, you must know who your readers are and what level of expertise they have attained with regard to the subject. Depending on what you know about your readers, you may have to define terms and explain unfamiliar concepts. Clarity is your goal, and that means explicitness. If, after you have considered your audience, you still are not sure what to include, mis-aim on the side of too much information. Too much information may inconvenience some readers; it may aggravate others, but it's a small price to pay. Too little information means the procedures can't be done.

Planning for procedures begins before you start to write. First, you have to know your subject. Find out how the subject came into being. If it is something that has been designed, new software, for example, find out why. What needs will it fill? Who will it be marketed to? You should examine the subject yourself. Play with it; use it. If it is something that must be assembled, assemble it. If it is a programming procedure that must be followed, follow it. See if it works. Find out its purposes, who will use it, and under what conditions. What are its good and bad features? Even though your company may not want you to say, "Here is where our product is likely to break or malfunction," you still need to know these things. You can deal with these problems tactfully in a trouble-shooting section.

If your subject is complex, divide it into parts. These parts will later become sections in short procedures or chapters in manuals. You may even have to subdivide the subdivisions. These divisions, marked by headings, enable readers to see the important parts of the procedure and how they interrelate. Naturally, how you divide material up depends greatly on the design you are using for the information. For example, modular design, which Edmond Weiss describes in detail in *How to Write Usable User Documentation*, requires that you subdivide information so that it fits into very specific sections of a document, usually no more than two facing pages and often only one page. Chapters and sections are less restrictive in terms of the physical requirements of document layout, but information in them must be highly organized, too.

If your subject lends itself to things that might go wrong when the procedures are not followed accurately, make a list of precautions your readers should know. The following are headings that have widely accepted meanings.

DANGER—reserved for steps in a procedure that could lead to serious injury or loss of life if readers do not know what they are doing.

WARNING—used for steps that could result in damage to the product if the procedures are not followed accurately.

CAUTION—applied to steps where faulty results could occur if the reader does not follow the procedure correctly.

NOTE—used to alert readers to potential problems.

ADVICE—reserved for making suggestions that would make the reader's work easier.

More and more companies are combining the last two, in a heading called COMMENT.

Once you have analyzed the subject, considered the projected audience, made an outline, and listed the precautions that will apply, you are ready to write the procedures.

WRITING PROCEDURES

Writing procedures begins with determining the purpose of the procedures for the intended audience. This includes choosing an explicit title that includes the date and all appropriate models and operations to which the procedures apply. It also often includes writing a short purpose statement introduction. This is the most often overlooked aspect of procedure writing. Many writers assume that since they are telling readers *what* to do, nothing is served by telling them *why* to do it. This is a dangerous fallacy because it assumes readers will automatically see the wisdom of doing it your way. Not so! Once these concerns have been met, you will have to choose a format that suits your purpose and communicates the information efficiently to your readers. There are a variety of choices. Figure 16 shows the VOS for procedures.

TESTING PROCEDURES

The final task procedure writers have to concern themselves with could be called quality control. Usually, we think of quality control as something that applies to production items, something that industrial engineers do. But quality control also applies to written documents, or information products, and it particularly applies to procedures.

To check the quality of your procedures, make sure that two types of people read and evaluate them before the procedures are made public. Have an expert in the technical area of the subject review the procedures for accuracy; this is often referred to as a technical edit. But don't stop there. Have a person who represents the ability level of the users

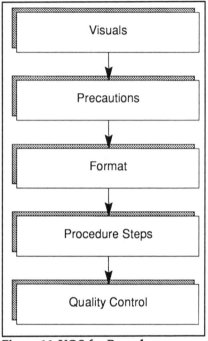

Figure 16. VOS for Procedures

read it to see if the procedure can be understood. This is known as a usability edit, and it is vital to the effectiveness of your final product.

After these edits, you might want to beta test the procedure in the environment in which it is intended to be used with persons intended to use it. This is a common practice for evaluating the functions of newly developed hardware and software. Since the first edition of this book was published, it has become increasingly common to extend beta test environments so that they include documentation. Some companies, such as IBM, have gone so far as to videotape users dealing with the hardware or software product and the documentation, making comments about their trials and tribulations as they go along. Aside from the fact that these videotapes are hilarious, they are extremely useful in creating improvements in both technological and informational products.

CONCLUSION

What I have done in this chapter is give a procedure for writing procedures. As is the case with all procedures, the success of this one will be determined on the basis of how well you are able to write procedures for a variety of audiences. Remember, however, that this is a procedure for developing a skill. Just as a potter, through practice, becomes competent at throwing pots, you, through just as much practice, will improve your procedures.

SUGGESTED READINGS

Beene, Lynn and Peter White, eds. *Solving Problems in Technical Writing.* New York: Oxford University Press, 1988.

Houp, Kenneth and Thomas Pearsall. *Reporting Technical Information.* 5th edition. New York: MacMillan, 1984.

Lunine, Leo R. "The Procedure Writer as a Catalyst for Change." *Technical Communication.* Vol.23, No. 4 (1976), pp. 10-11.

Weiss, Edmond. *How to Write Usable User Documentation.* 2nd Edition. Phoenix: The Oryx Press, 1991.

Whitehouse, Frank. *System Documentation.* London: Business Books, Ltd., 1973.

CHAPTER 16

How to Write Proposals

Strictly speaking, a proposal is a sales piece of writing, a communication designed to obtain work, funding, a "go-ahead" on a project, and so on. Its ultimate goal is to identify a need on the part of the audience and outline ways that the writer, or the group of people the writer represents, can satisfy that need. For proposals to be successful, writers must convince the audience that they can do something for the audience, that it needs to be done, and that the writers or proposing group can expect some sort of recompense for having done it.

As it appears, this is not easy. First, proposals must invite readership if they are to be successful. Readers must want to go beyond the first sentence of the introduction. That is one reason why three-part purpose statement introductions are particularly important to proposals. Such introductions get the reader oriented quickly to the aims of the document.

In addition, there are other ways to invite readership of proposals. Make sure that the proposal is easy to read and understand. Accordingly, you should try to do some, if not all, of the following.

- Consider the needs and level of understanding of the audience (i.e., analyze your audience).
- Use a simple format (more about that later).
- Make sure that the final draft is clear and legible (good letter quality print, attractive layout).
- Keep paragraphs and sentences reasonably short (shorter in proposals than in analysis reports, for example).
- Use headings (even in informal memo proposals).
- Use the active voice style whenever possible (more on that in a later chapter).

Finally, realize that readers read proposals in order to reject them. This is especially true of proposal evaluators in large, competitively bid projects. You will want to provide skimming cues for intuitive readers, as well as sufficient details about the project for sensing readers. Make it easy for people to judge your proposal by supporting it with logical reasons (for thinking evaluators) and with how it fits into their value systems (corporate, national, etc.) for feeling readers.

In this chapter, we will look at ways of writing successful proposals. Specifically, the following will be examined:

- proposal formats
- title pages for formal proposals
- proposal introductions
- discussion sections
- conclusion sections
- Gantt chart time schedules

Figure 17 depicts a method for organizing your proposals.

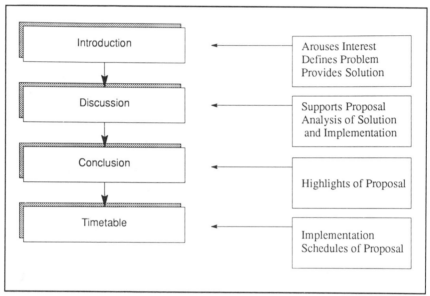

Figure 17. VOS for Proposals

PROPOSAL FORMATS

The following proposal formats are general guidelines, not prescriptions. More than likely, the company you work for (if it is a large company) has style guidelines for reports of this type. The material presented here, however, is easily fitted to your company's format. If your company does not have a style guide, the information presented in this chapter, and in fact in the rest of this book, can be used to help design a company style guide. In embarking on such a project, remember that the two most important aspects of report format design are the needs of the audience and the purpose of the writer and company.

All the formats presented in this section are for informal proposals, those without title pages, letters of transmittal, tables of contents, prefaces, acknowledgments, and the other trappings of formal reports. The following format is designed for problem-solving proposals.

Format for Problem-Solving Proposals

- Introduction—A three-part introduction that defines the problem to be dealt with by the proposal, briefly describes the proposed solution, and briefly outlines the benefits of adopting the solution.
- Discussion—An exhaustive analysis of the solution, including its benefits and challenges.
- Resources—What is needed to implement the solution, including what is available and what must be obtained.
- Costs—Dollars and cents (how much the solution will cost in dollars and cents?).
- Personnel—The people who will implement the solution and the people affected by it.
- Schedule—A Gantt chart for implementing the solution. (See p. 109 for a discussion of a Gantt chart.)

The following format could be called an executive summary proposal, since it aims to satisfy the needs of the reader who does not have the time or the inclination to read the entire report. The important criterion for a successful executive summary proposal is to include enough of the right kind of information in the summary for the reader to be convinced that you can do what you propose to do.

Format for Executive Summary Proposals

- Summary—A brief but complete discussion of what is at issue and what is being proposed. The idea is to give the executive reader enough information to make a decision at this point.
- Introduction—A transitional introduction that focuses on the background to what is being proposed and why. Orient readers here, preparing them for the detailed discussion that follows.
- Discussion—A detailed analysis of the proposal, its benefits, challenges, and implementation.

Note that this format is essentially two statements of the proposal—a "basics" version for the executive audience and a detailed version for the technical audience.

The final format to be considered here is an extension of the first two. Whereas either of the previous proposal formats could be used for memo proposals or short informal proposals, the following format is generally used for longer proposals, proposals that might be used to suggest a project of some sort. Formal proposals usually are based on some version of this format, as well.

Format for Project Proposals

- Introduction—A three-part purpose statement that orients the audience to the background which led to the development of the proposal, the specific goals of the proposal, and the purpose of the report in terms of its readers.
- Statement of the Proposal—A complete description of what is being proposed.
- Management Section—A comprehensive analysis of the budgetary and personnel responsibilities included in the proposal.
- Technical Section—A comprehensive analysis of the research, engineering, design, development, and material aspect of what is proposed.
- Costs—A detailed projection (often visual) of proposed costs.
- Schedule—A Gantt chart.

Note that in this format, writers often supply their own subject subheadings for shorter subsections within the major divisions of the proposal.

Some companies may require that proposals for large research and development projects be written as three separate but related documents:

- Technical proposal
- Management proposal
- Cost proposal

The proposals are merely an extension of the formal project format presented earlier. Appropriate front matter (title page, summary, introduction) and back matter (schedule, appendixes, resumes) would be added to the technical proposal, management proposal, or cost proposal depending upon which type of proposal was being written.

Realize, too, that these three elements of the proposal, whether they are written as one document or as three, make up the parts of the proposal that are most closely evaluated. In many cases, the proposal that offers the most or best for the least cost will be the "winning proposal." But in many more cases, deciding what is best is not a clear-cut decision for those who evaluate proposals. This is where the writer has considerable influence by making sure that the argument in the proposal is clear and persuasive. As I mentioned in the beginning of this chapter, proposal evaluators read proposals not to see where they are right but where they are wrong.

The next few sections will examine how to make proposals read right.

WRITING PROPOSALS

Title Pages

Title pages are often included on informal proposals and are always included on formal proposals. A title page should contain enough information so readers can easily indentify the contents of the proposal, its authors, issuing organization, and date of issue. It should be visually balanced, imparting a sense of order on the part of the reader. The not-so-subtle suggestion here is that the orderly appearance of a document reflects the orderly mind behind it.

Introductions

Introductions to proposals must orient readers to the contents and the purpose of the proposal, just as an introduction would do for any other report. The important difference, however, is the purpose of the writer. In most reports, writers are only relaying information. In proposals, they are *convincing* the audience that what is being suggested should be accepted, adopted, bought, or whatever. Persuasion becomes an important aspect of the introduction to a proposal.

Discussion Sections

In the discussion section of a proposal, writers describe the proposal, its benefits, and its implementation in full. Persuasion is no less important here. One way to enhance the readers' understanding of the discussion is to make sure that the organization of information follows some clearly identifiable pattern.

Depending on the format and the intentions of the proposal writer, the discussion section of a proposal may do any of the following.

- In problem-solving proposals it may present an in-depth description of the solution. This would include subsections on the benefits of the proposed solution, on any anticipated difficulties in implementing the solution, on the amount and type of work required to implement the solution, and so forth.
- It may present a detailed description of what was highlighted in the summary of executive summary proposals.
- It is usually subdivided in long project proposals into subsections centered around the managerial or technical responsibilities for what is proposed. Because of the increase in the amount of information communicated to the reader, the subdivisions are necessary.

The following is an excerpt from a problem-solving proposal discussion. Notice the clear organization of what is proposed followed by supporting statements.

HARDWARE PURCHASES

Unfortunately, we do not have a closet full of surplus computer hardware. However, the computers we do have will provide an excellent basis for further expansion.

As a result, I recommend the following:

1. That we no longer use cassettes to load computer programs.

2. That all computers be equipped with disc capability.

3. That we purchase a network system that will allow us to keep all our computers under one control.

4. That all future hardware purchases be compatible with our present computers.

5. That we offer inservice programs and summer workshops for teachers.

6. That we appoint a director of computer education.

I also recommend that we purchase the following:

1. One Radio Shack Network 2 Controller

2. Eight TRS Model III computers

3. Fourteen RS-232 connector cables.

The reasons for my recommendations are as follows:

1. If we upgrade to disc drive, then we will be able to buy better educational software.

2. A network system will allow a teacher to present the same skills to a complete class.

3. Radio Shack will install the equipment at no extra cost.

4. Radio Shack will give free inservice workshops to explain the equipment.

5. The new computers are compatible with our older models.

6. In order to develop curriculum, we must have workshops to put together a course of study and to order new software.

7. A director is needed to oversee the entire program. A director can order new software and be responsible for all the equipment.

Almost all the producers of quality software have switched from tape cassettes to disc memories. The advantages are many. The disc is less likely to lose information or to wear out from too much use. The disc has a much greater memory capacity, and the quality of the academic program is superior to that which can be obtained on tape cassettes. The tape cassette takes a much longer time to load, while information can be located on a disc almost immediately.

The network system, properly used, means that a teacher no longer has to run from one computer station to another station. The teacher can remain at the host computer, present the lesson, and download the lesson to the teaching computers. This will save considerable time.

Because of the compatibility of new Radio Shack hardware with our present computers, we need to buy only eight computers and one network system. This will give us a total of fourteen computers linked by a network. In addition, Radio Shack offers excellent services. They will deliver and set up all the equipment. Further, they will show us how to operate the hardware, and they will give us free workshops at the Framingham store. This is the only company which offers these extended services.

Finally, a full-time director will be able to put together loose details and direct future development. This is important because the classroom teacher does not have the time to participate in the running of a computer center and to maintain the responsibility for a classroom.

Conclusions

Conclusions to proposals serve two purposes: to put the reader in a positive, receptive frame of mind and to suggest to the reader how to obtain the benefits of the proposal. They should be brief and to the point. Benefits should be highlighted and required actions should be stated explicitly. Good conclusions give readers a way to assess the proposal, to see if the author is "on the right track." They also force authors to think beyond the proposing step, to examine the expected results of a successful proposal. By forcing the writer to think ahead, conclusions make time scheduling more straightforward.

Gantt Chart Time Schedule

All proposals should show the audience how long it will take to enact a successful project once it is adopted. One of the best ways of doing this is to use a graphic known as a Gantt chart. A Gantt chart takes a proposal and divides it into tasks to be accomplished and the dates by which they will be accomplished. It is a version of a common bar graph (explained in Chapter 7). Figure 18 is a sample Gantt chart of a documentation task. These charts can be used, however, for any process that allows tasks to be subdivided over a period of time.

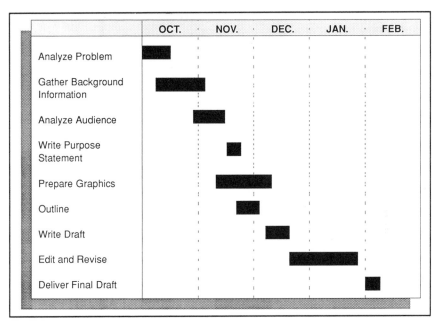

Figure 18. Gantt Chart

CONCLUSION

In this chapter, we have examined some of the aspects of writing successful proposals. Realize that there is no formula for guaranteeing successful proposals. Writing them takes a willingness to shape information so that it meets the readers' needs, so that it fits into the requirements of a company's style guide, so that it satisfies the author's purposes, and so that it is convincing to readers. This is a tall order, but it is not impossible. Just remember to make it as easy as possible for readers to evaluate and accept your proposals.

SUGGESTED READINGS

Beck, Clark E. "Evaluation of Research and Development Proposals." *Technical Communication*. Vol. 15, No. 2 (1968), pp. 11-14.

Clarke, Emerson. "Ten Steps to Better Engineering Proposals." *Product Engineering.* Vol. 36 (December 1965), p. 71.

Corey, Robert L. "Persuasive Technical Proposals: Rhetorical Form and the Writer." *Technical Communication.* Vol. 22, No. 4 (1975), pp. 2-5.

DeBakey, Lois. "Persuasive Proposal." *Journal of Technical Writing and Communication.* Vol. 6, No. 1 (1976), pp. 5-25.

Englebret, David. "Storyboarding—A Better Way of Planning and Writing Proposals." *IEEE Transactions on Professional Communication.* Vol. PC-15, No. 4 (1972), pp. 115-18.

Wexler, J. Ammon and Catherine Carmel. *How to Create a Winning Proposal.* Santa Cruz, CA: Mercury Communications, Inc., 1976.

CHAPTER 17

How to Write Analysis Reports

Analysis reports are called by different names from company to company—formal reports, project reports, final reports, job-end reports. I have chosen to call them analysis reports because that is the one characteristic they have in common: they all analyze a completed project to assess its success or failure. As such, these reports are an important aspect of documentation within the high-tech industries. Placed on file, they can be referred to by people working on the design of later products or by people writing user documentation.

In this chapter, we will look at a general approach to designing analysis reports. The important thing to remember, however, is that no report format is perfect. Company documentation standards attempt to resolve the issue by prescribing a format into which all analysis reports are poured. But even this does not work all the time. Report design should be flexible enough to meet a variety of writer purposes and audience needs. For example, you could consider fitting your analysis of a project into the overall value system of the corporation. Doing so could conform it to certain type preferences for maintaining order and past corporate values. Or, you could provide for a two-tiered readership by including summaries, headings, and overviews for skimming for intuitive readers and logically organized details for sensing readers. Since these reports are the basis of analysis, make sure that yours is rigorously logical. You will want to prove your judgments, especially for the thinking types in your audience.

The format presented in this chapter can be used as a basis for your own report design or as a starting point for developing report standards for your company. In addition, it can be used as a general guideline for organizing a formal technical paper since the difference between analysis reports and technical papers is often simply the medium of publication. Figure 19 shows this format.

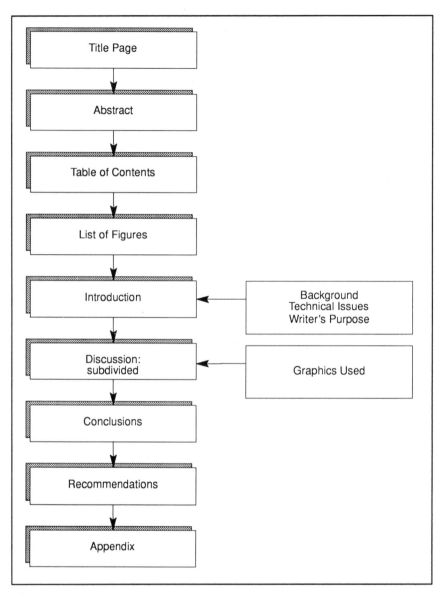

Figure 19. VOS for Analysis Reports

Each aspect of analysis reports will be discussed in detail, with the emphasis on options available to the writer.

SECTIONS OF A REPORT

Title Page

A title page should be designed with visual order in mind. It should be balanced from top to bottom and from left to right. It should provide enough information so that readers can tell what the context of the report is and what the report is about.

Titles should be kept relatively short, no more than 10-12 words. Avoid one- or two-word titles, however; they suggest that you are going to tell everything there is to know about the subject. And in most cases, that is not what you intend to do. Two-part titles can be a little longer. They are useful because the first part is general, providing the context; the second part is specific, indicating the main point of the report or the most important aspect of what was done. The following is an example of a two-part title:

PLOTFILE: A Software Interface for Accurately Graphing Large Data Sets

Abstracts

Abstracts are condensations of entire reports, focusing on the main issues: what was done, what was found out, and its significance. Abstracts are self-sufficient. Their use in many companies involves taking the abstract from the analysis report, photocopying it, and circulating it through a distribution list to readers who will then decide whether they need to order the complete report. This places a lot of responsibility on abstracts. They have to work for busy readers who do not have time to read the entire report and for researchers who want to know if the report is useful. To do these things well, abstracts must be informational.

Informational abstracts relate important aspects of the report. They should not contain words such as "described" or "presented." If they do, they are likely to be descriptive abstracts (little more than prose tables of contents) that tell what the report is about, not what information is important.

The following is an example of an informative abstract:

PLOTFILE is a program written for Concord Sciences Corporation to solve their graphing problem. Specifications were made by the president of the company and followed with two exceptions. The program now in use also has additional features. The user provides information through interactive commands on a video display terminal to specify desired plotting formats. With flexible logarithmic and linear capabilities, the program then draws the desired axes or grids and plots the set(s) of points read from the data file.

Notice that the important features of the program are highlighted in this abstract. Readers need to know that this information is in the report. Compare this informative abstract with the following example of a descriptive abstract.

This user's guide for the Concourse BASIC Interpreter describes commands, statements, functions, and the use of BASIC. Concourse BASIC is compatible with APPLESOFT BASIC and runs on PDP-11 UNIX systems.

You can tell what the manual is *about*, but you can't tell if there is information you can actually use.

Table of Contents

The table of contents provides an outline of analysis reports for readers who do not wish to read the entire report or flip through it looking for the section that contains what they are looking for. It should be made up of the headings and subheadings of the report, word-for-word with the accompanying page numbers. It should also be orderly and easy to read. The following is an example of a table of contents from an analysis report.

Page

Notice that in this example the appendixes are not numbered; in some companies the style guide requires that you number them. This is just one of almost countless small differences among company style guides.

List of Figures and List of Tables. These are separate tables of contents for all the figures and all the tables in an analysis report. Each is made up of the figure or table number, figure or table title, and number of the page on which the figure or table appears.

List of Symbols. This is an optional addition to the front matter of an analysis report; include it if you think some readers will need to have symbols defined. The same thing applies to the inclusion of a glossary.

Introduction

This is the place for the three-part purpose statement introduction. It will orient readers to the main issue of the report, to the technical issues or specifics that are important to the report, and to what the report is intended to accomplish for the readers.

Discussion

The discussion contains an analysis of the technical issues important to the report. It supports the main issue of the report by providing evidence and explanations. It should be subdivided into topics, each with a subheading.

Conclusions

This section presents the results of the analysis, the evaluation of what was presented in the discussion. Sometimes listing the conclusions is a good, concise way to organize them. It calls attention to the conclusions individually, but still enables writers to explain them as is necessary.

Recommendations

Not all analysis reports have recommendations, but if they do, the recommendations tell the reader what to do with the information provided in the report.

Appendixes

The appendixes are not a dumping ground for anything left over. They should serve a precise purpose. For example, if there is information that is not vital to the understanding of the main and technical issues of an analysis report, but that is still important to groups of readers, put it in an appendix. Such information often includes derivations of equations, tables of raw data, sample code, and so forth. But the only way to be certain that what is placed in the appendix belongs there is to assess it within the context of audience needs.

CONCLUSION

In this chapter, we have briefly examined analysis reports. The reason the examination was brief is not that these reports are comparatively unimportant. They are vitally important to companies. However, most of what was presented in the chapters about discussions and conclusions to articles can be applied to analysis reports. Most articles published in computer trade journals are analytical.

SUGGESTED READINGS

Beene, Lynn and Peter White, eds. *Solving Problems in Technical Writing.* New York: Oxford University Press, 1988.

Emerson, Frances. *Technical Writing.* 2nd edition. Boston: Houghton Mifflin, 1990.

Feinberg, Susan. *Components of Technical Writing.* New York: Holt, Rinehart, and Winston, 1989.

Van Duyn, Joan. *The DP Professional's Guide to Writing Effective Technical Communication.* New York: Wiley-Interscience, 1982.

CHAPTER 18

How to Write Product Descriptions

Descriptive techniques are primarily used for writing product descriptions. But product descriptions not only are used to describe a product, they also are used to sell the product or to attract a reader's attention to it. Many companies use product descriptions as a part of their public relations campaigns or as part of their advertising strategy. In large companies, these documents are often written by specialists, but in small companies *you* might be the specialist.

AUDIENCE NEEDS

The first thing to consider is what the audience needs from a product description. Use the following checklist as a guideline.

1. What is the product? Notice that this is a definition question.

2. What is the product used for?

3. What does the product do?

4. How does the product do it?

5. What happens after the product does it? One of the sources of failure in a product description is when the writer does not realize the reader's need to know when the description is complete. Answering this question meets that need.

6. What is the product made of?

7. What are the product's basic parts?

8. How are the parts related to make the product do what it does?

You'll notice that the last three questions are those which you are likely to think of first. I listed them last for precisely that reason. In order for those questions to be answered effectively in product descriptions, you must first answer questions 1-5.

For example, consider questions 2 and 3. These questions focus you on the product's purpose. All products (with the possible exception of pet rocks) have a purpose. Communicating that to readers aids them in understanding the product itself. Does your product have a single purpose or many possible uses? Notice here that I will equate purpose with use; they are similar and in product descriptions they are one of the most important details. Does the product operate independently or in conjunction with something else? Who uses it? And under what conditions? A former student of mine at MIT wrote a brilliant product description of a welding device he and a classmate had designed. The description was well organized, and it read well. But it did not explain who was supposed to use the product and under what conditions. I later found out that this particular welding device was intended to be used by remote control under the ocean up to depths of 3000 feet. It would attach bolts to pieces of sunken ships so that cable could be looped around the bolts and the ship salvaged. One could certainly argue (and I'm sure the U.S. Patent Office would) that this omitted detail was the most important aspect of the welding device. Don't overlook these sorts of details in your own product descriptions.

Treating issues of this sort makes your description all the more successful when you get around to describing the product's size, shape, and dimensions. When you begin to describe these physical characteristics, take care to orient the product in the reader's eye. Obviously, a photograph or line art will do a good job of this, but not all product descriptions have that luxury, due to budget restrictions. Realize that all products have a natural orientation, a side or view that a user would approach first. You will want to describe the product from that point of view. For example, how do you "see" a chair in your mind? I would bet you see the front of it, facing you, inviting you to sit down, That would be the view I would use to describe a chair. Think of how difficult a time an audience, which is not familiar with chairs, would have if you described it upside-down, or with the back facing the user.

Finally, don't forget the product's color and finish—if they are important aspects of the product itself. Color is important for fire trucks; it probably isn't for disk drives. Texture is important for nonskid surfaces, such as accelerator pedals, it probably isn't for cathode ray tubes.

Product Description Principles

For product descriptions to succeed, you must keep your purpose and intended audience clear at all times. These two issues govern the extent of details used in the description, the kind of details used, and the order in which those details are used.

Beyond that, product descriptions also use three types of details:

- those which describe the product's function or use
- those which describe the product's physical characteristics, and
- those which describe the product's parts or components

Finally, product descriptions can be general (for unsophisticated audiences who are unfamiliar with the class of products) or they can be specific (for sophisticated audiences who already know what the line of product is). You are more likely to encounter the first type of audience when a product description is part of a longer document, an owner's manual for example. The second type of audience is most often reached by stand-alone product descriptions used in public relations announcements or in advertising campaigns. Each type of description has its own format.

FORMAT FOR A GENERAL PRODUCT DESCRIPTION

Begin by identifying the product, usually with a definition. Explain why the product is important to your readers. Forecast where your emphasis will be—on function, physical characteristics, parts, or some combination of each. Think of this part of a product description as your introduction. In the body of the description, describe the functions, physical characteristics, and parts. Any order of these details will work, but make sure it matches your purpose for the description.

If you begin the body with a description of the product's function, be sure that you focus on who uses the product and when, where, and how they use it. If the product is used in conjunction with some larger system, don't forget to describe the functional relationship with that system.

When describing physical characteristics, try to enable your readers to "see" the product. Focus on characteristics that appeal to the readers' five senses, obviously choosing those that are important to the product itself. The aroma of natural gas is important when describing a stove, but the aroma of a microchip probably isn't.

When describing the parts of a product, first list them and then describe them in the order of the list. For each part, you will want to define it, describe its function, its physical characteristics, its relationship to other parts, and if necessary, subdivide it and run through the same sequence of material for all the sub-components.

Notice what is occurring here. I am suggesting that for general product descriptions you use a format that replicates itself. It is a nested format. For complicated products, you simply keep subdividing as often as necessary, repeating the description format at each level. Theoretically, I suppose one could eventually arrive at the sub-atomic level in a product description. Fortunately, I've never seen that done.

General product descriptions close by showing how the components work together to make the product do what it does or by mentioning unique features of the product or variations of it.

FORMAT FOR A SPECIFIC PRODUCT DESCRIPTION

One might think that for a specific product description the format would be even more involved and complex than for a general description. But that is not the case. You can safely assume that the audience of specific product descriptions shares more information with you than the audience of general product descriptions. As a result, specific product descriptions are shorter, tighter versions of general product descriptions.

For example, in the introduction to a specific product description you would want to combine product definition and function descriptions. The body of a specific product description is limited to a description of the product's features.

Close a specific product description with any of the following:

- the last feature of the product
- a summary of the product's benefits and uses, or
- a description of product variations

For one of the best models of specific product descriptions (and a model that is easy to obtain), go to your nearest automobile dealership and obtain a new car brochure. You'll notice that the cover describes, usually with a photograph, what the product is and the glamorous ways it can be used. The inside of the brochure describes the features of the product. And the back page lists all the options you can get. Obviously, new car brochures are more "glitzy" than most computer product descriptions, but that is not a generalization that holds across the board. Some new computer products are introduced with product descriptions that are every bit as flashy as a new car brochure. Naturally, such descriptions are the result of large budgets. If your budget is not that flexible, consider the following example from a start-up company I worked for a decade ago. The purpose of this description was to introduce a new product to the microcomputer and low-end minicomputer market. The writer wanted first to summarize (in a list) the features that were distinctive. Following that, the description defines SIGPAK and elaborates on its features. In this example, the writer opted for the closing to be the last feature.

Example of a Product Description

SIGPAK
An Intelligent Signal Processing Package for VENIX/UNIX™

* Signal Processing Package for Speech Analysis, Biomedical Processing, Seismics, and Psychophysics

* Features a powerful *Waveform Editor (WED)* modeled after the popular EMACS text editor, plus other standard signal processing utilities

* Provides a carefully thought-out set of *Data Structures, Protocols,* and *Runtime Routines* for custom program development

* Optimized for fast response on microcomputers and small minicomputers

SigPak is a UNIX-based package of *signal processing programs, runtime libraries, predefined data structures and protocols. SigPak* is a sophisticated tool for scientists, engineers, and programmers—anyone who needs to create, process, or manipulate signals. *SigPak* is specifically optimized for signal processing work on minicomputers, and includes a *waveform editor* called *WED* that optimizes waveform display and editing on a variety of medium resolution graphics terminals.

Using *WED*, interesting signal events can be located and labeled with text strings which are then stored in label files associated with the signal. *WED* can display multiple signal channels simultaneously. The command language of *WED* is based on the popular EMACS family of text editors. Just as a text editor can delete portions of a text file, and perform "cut and paste" operations, *WED* can cut and paste signal segments—and with approximately the same response time typical of text editors.

In addition to its other capabilities, *WED* constitutes a "shell" from which standard *SigPak* signal processing programs, and custom-designed user programs, can be accessed in a *signal calculator* mode. Signals can be recorded from a variety of standard signal sources (such as A/D convertors), and can be played back in real time. Signals can be added together, filtered, scaled, interpolated, and so forth. Fourier transforms can be calculated, and a variety of speech-signal-specific manipulations performed.

In addition to the prepackaged facilities of *SigPak*, which makes it possible to perform extensive manipulations on the signals without writing any programs, *SigPak* is specifically designed to assist the C programmer who needs to write digital signal processing routines. *SigPak* provides a carefully thought-out set of data structures, protocols, and runtime routines for custom program development. A flexible, efficient format for signal data files and signal parameter files is defined; and, a set of file I/O routines is provided for opening signal files and defining and moving "windows" through the signals. Indeed, the concept of "editing" a signal is consistently carried through the entire *SigPak* I/O library.

As mentioned before, a method is provided for labeling signal times and individual samples with text or numerical labels. *Signal parameters* can be defined and assigned values. These parameters are structured symbols that are created, stored, and updated by *SigPak* programs and by the user as a means of communicating between separate programs. In fact *SigPak* regards a signal primarily as a structured list of symbol names and values.

A standard set of symbols, including such items as sampling period and signal length, are automatically created and set when a signal is generated. At one level of interaction *SigPak* signal processing programs communicate using expression trees strongly reminiscent of LISP expressions, but optimized for use in a C and UNIX environment. Thus *SigPak* forms the basis for a *knowledge-based approach to signal processing.*

Finally, *SigPak* is designed so that the power and flexibility of the UNIX operating system can be used to best advantage. Label files and signal symbol files are text files that can be sorted and searched using *sort* and *grep*. Shell scripts can be written to manipulate signal files. Graphics routines make use of the standard UNIX plotting package. *SigPak* signal processing programs may be *efficiently* combined using pipes.

™VENIX is a trademark of VenturCom, Inc.
†UNIX is a trademark of Bell Laboratories.

PART 5

Finishing Your Work

CHAPTER 19

How to Avoid Common Writing Problems

Writing is one of the most difficult tasks anyone in any profession has to do. But this isn't news, is it? That is one of the reasons you are reading this book—to learn ways to make writing easier. What makes writing difficult is that it is creative, and creative processes take time. It is also hard work. Now, the notion that technical writing is creative may be foreign to you. But it is. You must create ideas to be communicated, words to carry those ideas, sentences and paragraphs that will be clear to readers, and overall organizational schemes that will make the logic of your ideas apparent. No wonder it's difficult.

Among the general difficulties of writing, however, are some common problems that occur regularly and that can be avoided. These are writer's block, organizational problems, punctuation problems, readability problems, and style problems. Every writer who has ever written—from the best known novelist to you and me—has had to work through these problems. For some, using trial and error, it takes a long time, and it is frustrating work. For others, using proven successful methods, it still takes time, but not as much of it. And it is no longer hopelessly frustrating.

WRITER'S BLOCK

Writer's block is what we call the experience of getting stuck while writing. Everyone has experienced it, so you're not alone. The writer writes and all of a sudden can go no further. Minutes and more minutes pass, sometimes adding up to days, and nothing gets done. Deadlines approach, and with them panic. Although the results are the same, there are several sources for writer's block: lack of information, lack of a well-defined purpose, lack of a thesis, poorly analyzed audience, and lack of confidence. Fortunately, each of these has a solution.

Lack of Information

The solution to a lack of information is simple: get more of it. But naturally, it's not quite as easy as that. First, one has to identify the problem. The surest clue is if you find yourself writing in circles, being repetitious, and not getting anywhere.

After you have decided that you do not have enough information or the right kind of information, you will have to discover the cause. Is it because you have not researched your subject thoroughly, or is it because you have kept poor records of your information? If it is the result of faulty research, then you will have to stop writing and do more information gathering. If you find yourself doing this often, then you might correctly guess that you have a problem in defining your communication purpose. More about this later. If your lack of information is the result of poor records, then you will have to redo some of the research in order to refresh your memory. The lesson of this is clear: don't trust your memory—at all, with anything. If it is important, or if it seems it may ever be important, *write it down.*

Regardless of the causes of your lack of information, when you discover that as the source of writer's block, you must stop writing. That's right; stop writing. Continuing to write at this point will accomplish nothing positive, and you will have to rewrite it anyway. After you have found enough information, then you can begin writing again, with a new, clearer sense of purpose. The important skill to develop is the ability to notice that a lack of information is the source of your writer's block. Remember that being repetitious is often the clue.

Lack of a Well-Defined Purpose

A poorly defined purpose for your communication will inevitably make writing more difficult. It may block it altogether. Sometimes, you will experience this as a lack of information; often you will notice it in paragraphs, sections, or entire reports that shift topic in the middle. For any case of poorly defined purpose, the solution is simple. Write the following sentence: "The purpose of this document is _____." And fill in the blank. Every technical document has a purpose, and all the information you include in the document should advance that purpose. If you want to make your work even easier, write the purpose statement sentence *before you begin to write* the document. Then you can develop a three-part purpose statement introduction around that sentence, as well as outline your report making sure that every entry relates to your purpose.

Lack of a Thesis

"Thesis" is simply communication jargon for "main point," the single most important point of information you want your readers to know as a result of reading your document. And every technical document has one of these, too. If you don't know what yours is, you are doomed to write rambling communications that will (perhaps) eventually get to a point. Of course, no one can use these things, so you're better off deciding what your main point is before you write. Then you can use all your information to develop it. The way you identify your main point is exactly the same way as you identify your purpose. Write the following sentence: "The main point of this document is _____." And fill in the blank. Remember, though, that your main point is not your goal; you identified that as your purpose. The main point is what you want readers to know.

Poorly Analyzed Audience

Although audience analysis was already discussed in Chapter 3, it bears repeating that a poor grasp of your audience can be a source of writer's block. You can recognize this best through experience. You find yourself staring at your computer wondering who is going to use your report and what on earth do they want from it. If that ever happens, recognize that these two questions are vital audience analysis questions. They should be answered *before you start writing*.

So, the solution to a poorly analyzed audience is also to stop writing and go do the audience analysis. Only after having done it will you be less likely to waste time doing nothing.

Lack of Confidence

A lack of confidence in your abilities as a writer is a common source of writer's block, and it is usually self-fulfilling. If you do not think you can communicate effectively, then you will not be able to do so. Practice of systematic approaches to writing problems is a solution to this problem. This also tends to be self-fulfilling. The more you follow a writing process that works, the more positive feedback you receive about your reports and papers, and consequently the more confidence you will develop in your communication abilities.

Writer's block, however, may also come from outside pressures. Recent research into writer's block suggests that much of it stems from unrealistic expectations about writing and unrealistic demands placed on writers. One of the most common unrealistic expectations about writing is that it can be done quickly. It can't, not even by intuitive-perceptive types who successfully complete writing projects at the last moment. As pointed out earlier, these types work long and hard turning ideas over in their minds, evaluating and discarding hypotheses long before they began to produce a document. Writing is arduous work, no matter how you prefer to do it. Because of the differences in our preferences for how we order our work, writing projects should be kept as flexible as possible. Certainly deadlines are a reality, but periodic milestones may even contribute to writers' block for certain people.

One other thing: if you find that you are stuck, whatever you do, don't continue to write. You will accomplish nothing. This may sound like advice to proscrastinate, and if abused, it is. But if you encounter writer's block, stop writing for the time being, and work to solve the cause of your block. You will waste less time over the course of writing a report or paper, and eventually you will develop ways to avoid the common causes of writer's block.

ORGANIZATIONAL PROBLEMS

Organizational problems are discovered during the editing phase. This is what makes editing vitally important to the production of technical documents. You want to ensure that organizational problems are found and fixed. If they are not, readers will experience them as complete breakdowns in the communication. They will have to puzzle out how you got from one point of information to another. No reader likes to do this, and most will not.

The solution to organizational problems has already been discussed at length in Chapter 9.

PUNCTUATION PROBLEMS

Accurate punctuation does not ensure accurate communication, but accurate communication is tremendously enhanced by accurate punctuation. Take the following punctuation test, without first looking at the corrected version that follows, to see how well you know punctuation. The test focuses on the most common punctuational problems found in technical writing. When you finish, compare your results to the corrected version and give yourself 1 point for every point of punctuation that agrees. If you score 40 or better, you should not worry about your punctuational abilities. If you score less than 40, the punctuation problems depicted in this test are discussed later in the chapter.

Punctuation Test

Punctuate the following sentences. Do not change anything else, such as the capitalization or wording. Two of the sentences require you to choose the correct word.

1. The report was bad and the presentation was worse.

2. Five reports are required a proposal a set of instructions an interim report a presentation and an analytical report.

3. The programmers John Jones Erin Davidson and Will Watson were well qualified for the job.

4. Because the proposal was late the company lost the contract.

5. If we think we can find a solution.

6. In that case let's submit a proposal.

7. Since he published "The Future of Artificial Intelligence Professor Ward has become famous.

8. Professor Ward wrote The Future of Artificial Intelligence

9. Professor Ward wrote The Future of Artificial Intelligence he has become famous as a result.

10. Walters report Asynchronous I/O in Minicomputers was well received.

11. Its well known that the groups findings contradict its assumptions.

12. If we look at the diagram Figure 4 we can see the problem.

13. None of the reports (was/were) correct in (its/their) assumption.

14. A phoneme an element of sound is one of the building blocks of spoken language.

15. The finding of the seven different reports on extraterrestrial intelligences (was/were) made public today.

16. The reason our proposal was rejected is simple it was too expensive.

17. The manager rejected our proposal for the following reasons it was too expensive our conclusions were vague and the time schedule was too long.

18. Who wrote The Future of Artificial Intelligence

19. The feasibility study was inconclusive however the project will be done anyway.

20. The reports findings indicate that because the procedure was incorrectly followed it is impossible to fund the full scale project.

Corrected Version of the Test

1. The report was bad, and the presentation was worse.

2. Five reports are required: a proposal, a set of instructions, an interim report, a presentation, and an analytical report.

3. The programmers—John Jones, Erin Davidson, and Will Watson—were well qualified for the job.

4. Because the proposal was late, the company lost the contract.

5. If we think, we can find a solution.

6. In that case let's submit a proposal.

7. Since he published "The Future of Artificial Intelligence," Professor Ward has become famous.

8. Professor Ward wrote "The Future of Artificial Intelligence."

9. Professor Ward wrote "The Future of Artificial Intelligence"; he has become famous as a result.

10. Walter's report, "Asynchronous I/O in Minicomputers," was well received.

11. It's well known that the group's findings contradict its assumptions.

12. If we look at the diagram (Figure 4), we can see the problem.

13. None of the reports was correct in its assumptions.

14. A phoneme, an element of sound, is one of the building blocks of spoken language.

15. The finding of the seven different reports on extraterrestrial intelligences was made public today.

16. The reason our proposal was rejected is simple: it was too expensive.

17. The manager rejected our proposal for the following reasons: it was too expensive; our conclusions were vague; and the time schedule was too long.

18. Who wrote "The Future of Artificial Intelligence"?

19. The feasibility study was inconclusive; however, the project will be done anyway.

20. The report's findings indicate that, because the procedure was incorrectly followed, it is impossible to fund the full-scale project.

This test covers all the common punctuation problems (and some of the grammatical problems) writers are likely to face. What is not covered here can easily be found in a good handbook on punctuation. One of the misconceptions about punctuation, and language use in general, is that there are a thousand rules with ten thousand exceptions. This is not exactly accurate. There are some rules, and there are some exceptions, but not enough of each to preclude being able to remember them. The rest of this section will explain what the test covered. After you have compared your answers with the correct ones, you can see how much or little you need to commit to memory.

Sentence 1 exhibits a common problem concerning punctuation: the use of commas. When two sentences are joined with any of these words (and, or, so, but, for, yet, nor), a comma is always required to separate the sentences.

Sentence 2 deals with the use of colons (:). Colons are used to set off lists from the rest of a sentence. Commas are used to separate the items of the list. The comma before "and" is optional. But leaving it out often leads to confusion and misreading. Put it in.

Sentence 3 displays ways to insert nonessential material into sentences. Usually, writers can do this in any one of three ways: with commas, with dashes, or with parentheses. In this case, however, the material that is inserted is in the form of a list. Because it contains commas, you are required to use either dashes or parentheses to avoid confusion.

Sentences 4, 5, and 6 deal with introductory material placed at the beginning of the sentences. If the introductory material itself contains a sentence after the transitional word ("Because" or "If"), a comma is required. Otherwise, as in the case of sentence 6, no comma is necessary unless you want the reader to experience a definite pause. Placing the comma in or leaving it out has nothing to do with the length of the introductory material.

The important thing about sentences 7, 8, 9, and 10 is the treatment of quotation marks in relation to other marks of punctuation. Periods and commas are *always* placed inside quotation marks. I know some of you are saying, "But that's not the way I was taught." It's not the way I was taught, either, but the English punctuation system is dynamic. It changes. And the changes are toward greater simplification. Semi-colons are placed inside the quotation mark if they are part of the material being quoted (for instance, a title); otherwise, they are placed after the quotation mark. The same thing holds true for colons, dashes, question marks, and exclamation marks.

Sentence 11 exhibits the common problem of what to do with apostrophes. It seems most people ignore them, and this sentence is an example of what can come of doing that. The first word of this sentence is a contraction of "it is." The apostrophe is *always* included. The word "group's" is a possessive noun. The apostrophe is included here, too, to show that something belongs to the group instead of there being more than one group, a plural. The next-to-the-last word in the sentence is the possessive form of the pronoun "it." I realize this is confusing, but the possessive form of "it" *never* has an apostrophe. This is something which all of us simply have to commit to memory.

Sentence 12 is a version of sentences 4 and 5. The difference is the inclusion of "Figure 4." If you place the parentheses around this term, you have to lift the comma and place it *after* the final parenthesis to show that the term goes with the introductory material and not with the rest of the sentence. Otherwise, the sentence will not make sense.

Sentence 13 depicts a source of confusion that has been in existence for two generations. It probably confused a number of you. The logically correct version of the test lists a form that sounds odd. The reason it does is that this is an example of the difference between conversational English and written English. Most people are no longer offended by "None were" in conversation. Increasingly, usage experts are saying that it is acceptable in written language, as well. Twenty-five years ago, it would have been thought an example of poor teaching, however. "None was" is both correct and logical. "None" is the contracted version of "not one." This is an example of our language in flux. People who are more conservative in their approach to language use continue to use "none was." Journalists are a good example of this. More and more people, however, are comfortable with "none were." And most of us will live to see it universally accepted over the next generation.

Sentence 14 is identical to sentence 3. Here, however, all the options are available to you. Be sure you choose among them for the right reasons, though. Communication research has suggested that dashes call more attention to the material between them than commas and that parentheses call less.

Sentence 15 is an example of how easy it is to be careless when writing. The subject of this sentence, "finding," is separated from the verb, "was," by eight words. Writers can easily forget the number of the subject over that span. This is one of the reasons that editing is important.

Sentence 16 exhibits a different use of the colons. Colons can be used to signal to the reader that information is coming up which will explain information that has just been read. This sentence and sentence 17 are examples of that.

Sentence 18 deals with coordinating quotation marks with other marks of punctuation. See the explanation of sentences 7, 8, 9, and 10.

Sentence 19 treats the punctuation of a different type of sentence joining word. When sentences are joined with words such as "however," "therefore," "consequently," and about forty other similar words, the word is preceded with a semi-colon and followed by a comma. This signifies the longer or stronger pause these words require.

Sentence 20 also deals with punctuating material that is additive to the sentence. Here, however, the punctuation is entirely dependent upon whether or not the writer wants the reader to feel a pause. If not, it can be omitted entirely. One other aspect of this sentence is the hyphenation of "full-scale." Whenever two words are used together to modify another word, they become what is known as unit modifiers and require hyphenation.

This little exercise should let you know where you stand with regard to punctuation. If you did well, terrific. If not, solving the problem will not require as much work as you might have thought. See the suggested readings at the end of this chapter.

READABILITY PROBLEMS

Readability is a buzz word that most of us have heard. Much has been said about it, and there are numerous formulas that supposedly test it. Readability is the likelihood that a projected audience will be able to read and comprehend a piece of documentation. It is an extremely important aspect of writing in any technical field, but particularly in the computer industry. There is a problem, however: no one has agreed yet on a readability formula that works.

Let's look at two popular and widely used formulas, examining their strengths and weaknesses.

Gunning's Fog Index

This is a simple formula aimed at locating the audience on a grade scale that is supposedly based on their formal education.

1. Select a segment of text that is approximately 100 words long, to the nearest period. For more accurate results, choose text from the middle of a document. Introductions, leads, and conclusions usually exhibit slightly different styles that may skew the results of this test.

2. Count the number of sentences in the selected text.

3. Determine the average sentence length by dividing the number of words by the number of sentences.

4. Count the number of long words (those with 3 or more syllables). But don't count proper nouns, words that have 3 syllables because prefixes or suffixes have been added (for example, create-d), or words that are combinations of one- or two-syllable words (for example, storyboard).

5. Add the number of long words to the average sentence length.

6. Multiply this result by 0.4.

7. The result is the Fog Index. If you place it on a scale of 1-21, you will have the approximate number of years of formal education a reader would need in order to understand the document easily. 1-12 corresponds to grade and high school; 13-16 corresponds to college; 17-18 corresponds to a master's level college education; and 19-21 corresponds to a doctorate level.

NOTE: To increase the reliability of this formula, take several samplings from throughout a document.

Flesch Readability Scale

This test, like Gunning's Fog Index, is based on the length of words and sentences. It strives for greater accuracy by adding features that are analyzed, such as the total number of syllables. It also has the advantage of not being limited to a particular amount of text. Any size selection will do. In fact, several text analysis software packages include the Flesch Readability Scale as a provision to analyze entire documents.

1. Determine the following:
   a. total number of words (A)
   b. total number of sentences (B)
   c. total number of syllables (C)

2. Divide the total number of words by the total number of sentences to obtain the average sentence length (D)
   $A/B = D$

3. Multiply the result by 1.015.
   $D \times 1.015 = E$ (approximately 20)

4. Divide the number of syllables by the number of words to obtain the average word length.
   $C/A = F$

5. Multiply this result by 84.6.
   F x 84.6 = G (approximately 150)

6. Add E and G.
   E + G = H

7. Subtract H from 206.835.
   206.835 - H = Flesch Score

8. Place this score on the following scale.
   90-100—very easy
   80-90—easy
   70-80—fairly easy
   60-70—standard
   50-60—fairly difficult
   30-60—difficult
   0-30—very difficult

Although most explanations of the Flesch test also go into educational levels, such labels are not necessary. The scale does a fairly good job, as readability tests go, of explaining the relative difficulty of a document.

Both of these readability tests share a common fault. They oversimplify the task and the product of communication by ignoring the single most important issue—audience. Communication can not take place in a vacuum, and for that reason it is very difficult (some would say impossible) to measure its effectiveness quantitatively. In addition, these tests are based on word length and sentence length. This ignores the integrity of the subject matter. Try writing a few paragraphs about liberty, independence, declarations, revolutions, and constitutions. Your result will run afoul of each of these readability formulas.

While it is generally true that shorter words and shorter sentences are easier to read, it does not follow that they are always easier to understand. Sometimes, the short, choppy nature of such a style leads readers to skim a document, not paying close attention to what they are doing. Short sentences and simple words are best used in summaries, introductions, leads, and conclusions. A blanket indictment of longer sentences robs our language of its stylistic richness. It can even lead to stylistic contrivances. Writers are better off not having to worry about scoring below some arbitrary number on a less-than-effective scale. But writers *must* consider the level of expertise of their readers. This consideration is why audience analysis based on the knowledge shared between reader and writer is more important than the presumed reading level of that audience.

WRITING STYLE PROBLEMS

One's writing style is also an editing phase problem. At the draft phase, writers should be concerned about communicating subject matter. Style is secondary. But at the editing, polishing phase, style becomes a way of enhancing the communication and making it more accessible.

Style is the result of sentence structure and word choice. Although you might not think so, sentence structure is simple to understand and use. There are four basic types:

- Subject-verb (SV)
  The programmer (S) quit (V).
- Subject-verb-object (SVO)
  The programmer (S) kicked (V) the terminal (O).
- Subject-linking verb-complement (SLVC)
  The programmer (S) felt (LV) sick (C).
- Subject-verb-indirect object-object (SVIO)
  The programmer (S) gave (V) the engineer (I) a headache (O).

By itself, this is a lot of variety that writers can use in crafting a varied and interesting style. When we add to this basic structure ways to modify words and sentences, the possibilities become almost inexhaustible.

Simply put, sentences may be modified in three ways:

- Left-branching, or before the main part of the sentence
  When we add to this basic structure ways to modify words and sentences, the possibilities (S) become (LV) almost inexhaustible (C).
- Right-branching, or after the main part of the sentence
  Writers (S) should vary (V) sentence structure (O), allowing readers the opportunity to pause, digest what they have read, and go on.
- Mid-branching, or amid the main parts of the sentence
  Writers (S), if they are wise and experienced, vary (V) sentence struture (O).

It is the variety of sentence structures and lengths that make for a readable style, not the overuse of any one. Even though there are other ways to vary your writing style, these seven options can be used to create 46 different types of sentences. That's plenty for most of us.

SUGGESTED READING

Beene, Lynn and Peter White, eds. *Solving Problems in Technical Writing.* New York: Oxford University Press, 1988.
Flesch, Rudolf. *The Art of Plain Talk.* New York: Harper Brothers, 1946.

—————. *The Art of Readable Writing*. New York: Harper Brothers, 1949.

Gunning, Robert. *The Technique of Clear Writing*. New York: McGraw-Hill, 1952.

O'Rourke, John. *Writing for the Reader*. Maynard, MA: Digital Software Publications, 1976.

Williams, Joseph M. *Style*. New York: Scott-Foresman, 1985.

CHAPTER 20

How to Edit and Revise Your Work

Editing and revision are the final steps in the process of writing a report or article about high technology. These processes ensure successful communication. Almost everyone who writes does some sort of editing and revising, but many do not do so systematically with a specific goal in mind. For most, the idea of what constitutes a good report or article is only a gut reaction to their own work. Something seems good or something seems to need more work. While this process may succeed for some writers who have practiced it for a long time, it is difficult to develop the skills that are required to make accurate gut-reaction criticisms of one's own work. There is a better, easier way.

This chapter will present some ways that writers can assess the quality of their work and the work of others. It will also introduce the concept of an editing "buddy system."

DIVISION OF EDITING

Good editing divides the task of improving a document into a limited number of areas:

- organizational logic
- mechanical development of topics
- writer's style
- quality of the manuscript

Each of these areas is important to completing a successful, usable document.

Organizational Logic

Organizational logic is the single most important issue in writing a report or article about technology. It is the first thing a person needs to look for when editing a document. The ideas in a report or article should be interrelated—seamlessly. They should form a sequence of information that will appear to be predictable or inevitable to readers. In other words, a topic or an idea should anticipate topics or ideas that follow it. As I have said before, there should be no surprises in a document, no areas of confusion that the readers have to puzzle out for themselves. Edgar Allan Poe, referring to the writing of short stories, gave some advice to writers which we can very well borrow for the writing of reports and articles about high technology. He said that authors should include *nothing* that does not advance the topic towards its inevitable end. Anything else is a tangent that readers will wander down with the writer and become lost together.

Mechanical Development of Topics

How the writer develops a topic goes hand in hand with organizational logic. If the ideas are arranged in a logical order, development of the topic becomes a fairly easy task.

When editing an article, look at the lead first. Is it effective? Does it bring readers into the topic? Is it interesting, attention-getting? It has to be for the rest of the article to work.

When editing a report, look first at the introduction. Does it have three parts—background, a statement of the specific topic, and a statement of what the report will do for readers? If it doesn't, *rewrite it*!

For both reports and articles, look next at the middle. Does the discussion maintain and develop the topic as it was suggested in the lead or introduction? Is the overall progression of ideas predictable? Are there unnecessary surprises?

Examine the paragraphs individually. Each paragraph should begin with a topic sentence that gives readers an idea of what the topic of the paragraph will be and how it relates to the overall topic and/or purpose of the document. Then the paragraph should develop that topic and that topic only. Be on the alert for tangents in a paragraph, and remove them.

Look to see if there is transition between paragraphs, even between sentences. Transition is what links the ideas together. Transition can be accomplished in three ways. The writer might repeat an important word in two adjoining paragraphs or sentences, as I did with "transition" in the first two sentences of this paragraph. Structure can accomplish transition as it has in the way this paragraph and the three before it begin. Finally, transitional words and phrases establish links between ideas in an obvious way, as "Finally" does for this sentence and as "When editing a report" does for the third paragraph of this section.

Examine the ending of the report or article. Is it written with an obvious purpose in mind? Review Chapter 11 on exits and conclusions just to make sure that the purposes of conclusions are second nature to you. Will the ending leave readers with a sense of fullness and completion, a sense that you have satisfied their needs? It should.

At last, check the punctuation throughout the document. It should reflect the logical organization and interrelationship of ideas throughout the document. It should also be accurate and correct to presently accepted standards. If you do not know what those standards are, buying a good handbook on grammar and punctuation would be a wise investment. Even if you do have a good idea as to what society presently thinks is good punctuation, you still should have an accurate handbook for easy reference. I have one placed conveniently within reach of every place I write.

Writer's Style

Style is the writer's voice coming through the words and sentences. Style is inevitably structure, how writers arrange words and sentences in patterns that sound natural to them. Written style, however, differs slightly from conversation. It is a bit more formal, a bit more planned and measured. When we converse with someone, as opposed to when we make a planned presentation, there is little time to plan the arrangement of words and sentences so that what we say is clear. In conversation, we establish clarity by observing visual and oral cues from the other person to check whether we are being understood. But in writing, we do not have that luxury. We must anticipate readers' reactions, and we must be sure that the structure of our sentences and paragraphs represents the thoughts that are carried by them.

For example, are the sentences in a document varied in length and structure? If sentences are of nearly the same length and the same structure, the style will be monotonous and boring, regardless of the readers' interest in the topic. If necessary, review Chapter 19 on solving common writing problems to see how sentences can be varied.

Check your choice of words. Will the words mean for your readers what you intend them to mean? Are they words the readers are familiar with? Whatever you do, don't choose words to impress your readers. You won't. So if you regularly write with the thesaurus by your side, move away from it. Impress readers with the crystalline brilliance of your ideas, not with the ponderance of your vocabulary. If readers can't understand what you are trying to say, most readers will simply put your document down and go about what they see rightfully as more important work.

Is the style appropriate for the topic and for the audience? In this book, I have attempted a fairly direct, informal style to create a sense that I might be talking to you in one of the professional seminars I hold in industry. You, the audience, and the topic determined that. And you probably did not believe me when I said that readers were the most important aspect of communication. Readers have power. They can refuse to read! If this book had been written for professors of communication, or as a textbook for a technical writing class, the way it "sounds" would have been inevitably different, perhaps more formal and rigid.

Finally, does the style convey the effect the writer intended? If you wish to sell an idea or product, then the style must be persuasive. If you wish to convey information only, such as in a report to superiors, then the style should appear to be objective, even though real objectivity is a fiction.

Quality of the Manuscript

The last thing a writer needs to examine when editing a document is the quality of the manuscript. This is the icing on the cake, so to speak. It is the writer's last chance to polish the work.

A manuscript should be visually attractive. This means that it should be orderly, that there should be obvious divisions and subdivisions. One of the best ways to test your manuscript for this is to lay it out on the floor and stand over it. Can you see clearly where the sections are divided? Is there sufficient white space before and after headings? If you cannot tell, revision is called for.

A manuscript should invite readership. Have you ever seen a report or paper that had the print crowded out to the edges of the paper from top to bottom and from side to side? Such a manuscript *looks* hard to read. The type should be clear in a manuscript, and the production should not be sloppy. Avoid dot matrix printers, except for the most informal of interoffice memorandums. Final drafts should be more polished.

A report or paper should be sufficiently detailed. How much detail is sufficient detail? Only the writer can answer that accurately. There should be enough detail, though, to satisfy the projected readers' needs for information. Again, the answer is tied to audience analysis. Even though detail is vital, the document must be brief. These are not mutually exclusive criteria. If the writer has paid attention to logical organization and development of the topic, the document will be as detailed and as brief as it has to be.

After you have done all this, proofread the document—preferably in hard copy rather than on a screen. Don't confuse proofreading with editing and revision, however. These are much more involved processes, and proofreading is only one small step. Editing and revision require being judgmental, asking whether or not a certain statement or organization or figure is as good as it can be, and then making it so. Proofreading is a quick check for simple errors. The best way to keep them straight is to remember that editing includes proofreading but that proofreading alone does not ensure editing and revision.

THE EDITING BUDDY SYSTEM

So far in this chapter, we have examined what writers should do when editing and revising documents. Now let's look at how they should do it. The first thing you will want to do is to choose another person, your "buddy," whose critical judgment and honesty you can trust. You will agree with that person to edit each other's documents. One way to do this is to read your partner's document out loud to that partner. Every place you stumble, or hesitate unnaturally, signifies a place where the meaning is not as clear as it should be. Your partner makes note of this and rewrites the troubled area later. When you have a document that needs editing, your partner returns the favor. This is an extraverted procedure. If the physical arrangement of your workplace makes it impossible to edit a document by reading it aloud, you can accomplish the same results with the following procedure.

Write out on a separate sheet of paper a phrase or sentence which accurately describes the central point of each paragraph in your document. Give the document to your partner and have that person read it, doing the same thing for each paragraph. Compare the two lists. There *will* be differences. Discuss the points of difference with your partner. By doing so, you will gain insights as to how sections should be rewritten. This is an introverted procedure.

Finally, make sure that each document receives two types of edits: a technical edit, by an expert in the subject area, to ensure that the information is correct; and a usability edit, by someone who represents the audience's perspective, to ensure that the information can be understood.

CONCLUSION

In this chapter, we have looked at how writers can systematically edit and revise documents to ensure high quality. Editing and revision are vital, and the successful writer never overlooks them. In the high-tech industries, it seems that everything should have been done last week, or worse, last month. Nonetheless, take the extra day to edit and revise. The ends support the means.

To conclude this chapter, I've summarized the points of editing and revision in a checklist. Use it until it's memorized.

Editing Checklist

ORGANIZATIONAL LOGIC

1. Are ideas related?
2. Is the sequence of ideas clear?
3. Does it make sense?

MECHANICAL DEVELOPMENT OF TOPICS

1. Does the lead or introduction work?
2. Is it interesting?
3. Does it attract attention?
4. Does it include background, a statement of the topic, and a statement of what the document will do for readers?
5. Does the discussion or body develop the topic according to what was said in the discussion or lead?
6. Does each paragraph begin with a topic sentence?
7. Does the paragraph develop the stated topic?
8. Is there transition between paragraphs?
9. Is there transition between sentences?
10. Are the words accurately chosen?
11. Does the ending work?
12. Does it tie down the subject?
13. Is the punctuation accurate?

WRITER'S STYLE

1. Is the style appropriate for the topic and audience?
2. Is the style varied?

QUALITY OF THE MANUSCRIPT

1. Is the manuscript orderly?
2. Are there enough headings and subheadings?
3. Are the headings clear?
4. Is the print easy to read?

SUGGESTED READINGS

Bennet, John Bernard. *Editing for Engineers.* New York: Wiley-Interscience, 1970.

Boomhower, E. F. "Producing Good Technical Communication Requires Two Types of Editing." *Journal of Technical Writing and Communication.* Vol. 5, No. 4 (1975), pp. 277-81.

Price, Jonathan. *How to Write a Computer Manual: A Handbook of Software Documentation.* Menlo Park, CA: Benjamin/Cummings, 1984.

Williams, Joseph M. *Ten Lessons in Clarity and Grace.* 3rd Edition. Glenview, IL: Scott Foresman and Company, 1989.

PART 6

Presentations and Meetings

CHAPTER 21

How to Make Professional Presentations

Many professionals in high-tech industries find that a significant number of papers and reports turn into presentations, but presentations are different from written reports in some important ways. When people read reports, they can pay attention or not, because they can always reread. If some part of a report is unclear, they can go back over it more slowly, attempting to puzzle it out. These are not excuses for sloppy writing practices, but they are realistic advantages of written communication. When an audience is listening to a presentation, they do not have these advantages. They hear the message only once, provided that the speaker is not terribly repetitious. They have to focus on what is happening in the present, for if they ponder on what has just occurred, they are missing the present. They must pay close attention at all times because they cannot go back. They cannot skim or look ahead. And if they have questions, often they are unwilling to ask them (although it is easier to ask a speaker a question than to ask a book). If you have ever sat through a poor presentation, you have no doubt experienced some or all of these difficulties.

Because of these limitations, presentations must be flawlessly clear. Rigid organization is a must. The audience, in order to understand the message, must see how the speaker got from point A to point B, for all the points in the presentation. This is a big task, but fortunately there is a systematic way to approach it.

The system presented in this paper will not make your presentations spellbinding. The content and your personality have something to do with that. But it will ensure that your presentations are focused on an intended audience, that the information is clear and logically developed, and that you do nothing to distract from the message. In other words, this process for making presentations guarantees some level of success, but the polish and resulting accolades are up to you—the rewards of practice. Figure 20 depicts the process.

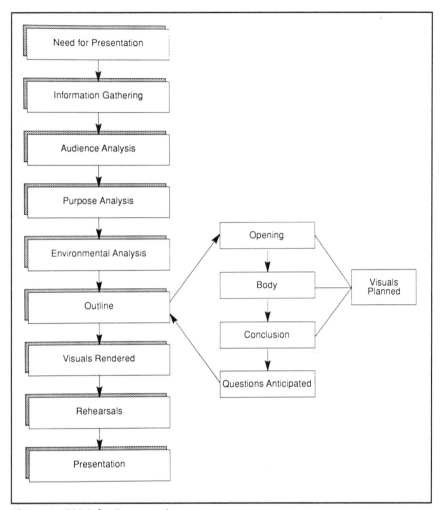

Figure 20. VOS for Presentations

PURPOSE OF THE PRESENTATION

The first thing you must decide is the general purpose of your presentation. There are only two choices—to inform or to persuade. Of course, most persuasive technical presentations are also informative, or at least they should be. However, deciding the general purpose of your presentation is by itself not much help. The important step is the next one. You should also decide on a specific purpose for the presentation. In other words, are you going to inform the audience about topic x, or are you going to persuade them to buy product y or to do project z?

Once you have decided this, consider whether the presentation might also afford you a personal purpose—such as advancement, fame, money, and so on. Does any personal purpose that might apply to the presentation complement or contradict your specific purpose? As an example, let's imagine that you are scheduled to make a formal presentation before the American Computing Machinery International Convention. During this presentation, you will announce the completion and successful beta testing of a new product line from your company. Such a presentation may serve to associate you with the new product line. This is undoubtedly an accurate association, otherwise you would not have been selected to make the presentation. Such an association might affect your career path in a positive way—either by advancement within your company or by the offer of a new position within another company. Anything of this nature would be personal purposes that complement the specific purpose of your presentation.

Let's imagine another situation. You have been chosen to make the public explanation to a group of large, consistent, corporate customers as to why a promised product is already months behind schedule and apparently not as good as your chief competitor's newly released product, an activity known widely as "walking the plank." You realize that your company, and by association you, will be seen negatively. This could cost your company millions of dollars in potential business; it could cost you your reputation. This type of situation demonstrates personal purposes that are in contradiction to the specific purpose of your presentation—to inform an audience about a problem and convince them that it is not as bad as it seems.

As these examples show, the determination of purpose is extremely important when organizing a formal presentation. It determines the information you will select, the points you will make and not make. Purpose determines what you will say.

THE PRESENTATION AUDIENCE

There are audience issues that differ from those concerning written reports. It is important to know what level of expertise the audience has attained in your subject area, as well as who they work for and what they do. But you should also consider the size of the audience. This tells you something about the size of the room in which the presentation will be held, and it limits the type of visuals you can use. (See Chapter 23.)

You should also consider these questions: What motivated the audience to attend your presentation? The subject matter? You? Were they required to attend? What specific objectives do they hope to realize by listening to you? The answers to these questions suggest ways to organize your information; they also suggest what should be included and what should be left out. For example, if your audience was coerced into attendance, you need to offer them something important or interesting early in your talk. If they chose to attend, seeking particular information, you can lead them to it a little more easily.

What does the audience know about you? Will they view you as credible or not? What information can you include that will enhance your credibility? These are extremely important questions if you are organizing a presentation before a large group at an industry conference. Realize that in such presentations you have about three minutes to convince the audience that you have something important to say before their attention drifts off, or in some instances before they physically leave. For this reason, the opening of a presentation is very important.

What does the audience know about the specific topic of your presentation? Do they hold any emotional ties (pro or con) to the subject? For example, speaking about nuclear power is almost guaranteed to polarize audiences into camps. Knowing or suspecting the attitudes of an audience is therefore important. You can anticipate problems and defuse them before they become serious. This is a skill that carries over very well into the organizing of meetings, where problems tend to be rife and attitudes strongly defended.

THE PRESENTATION ENVIRONMENT

Where the presentation is to be held is another issue to consider. If possible, examine the room. Is it a large auditorium or a small conference room? Will the audience be seated theater-style or around tables? This, too, affects the types of visuals you may use and your freedom of movement. It may even affect the formality of your presentation.

Are there potentially distracting elements in the environment such as lawnmowers running outside the window or humming fluorescent lights? Distracting elements can make it difficult for your audience to pay attention. Many distractions can be changed or avoided. If you encounter those which cannot, you will at least be prepared for them if you examine the environment prior to your presentation.

What will the arrangement be for you, the speaker? Will there be a podium, a microphone, a table, visual equipment? All of these questions should be answered before you arrive to make a presentation. Remember, you will be nervous anyway; don't let unnecessary surprises make it worse.

Is there a time limit for your presentation? The amount and type of preparation that goes into a ten-minute speech is vastly different than that required for a two-hour training presentation. And the ten-minute speech is more difficult! Will questions be allowed during the presentation or afterward? What will the audience be doing before and after the presentation?

These environment questions are certainly not as important as topic-related questions or purpose-related questions. But the environment for a presentation does affect its acceptance. Don't forget to consider it.

THE PRESENTATION TOPIC

There are some important areas to define concerning the topic of a presentation. For example, what do you know about the topic? And closely related to that: what do you need to know in order to make an effective presentation for your chosen audience?

Once you have gathered all the information you need to meet your specific purpose, you will have to organize it. Determine a limited number of main points to be made during the presentation, usually no more than five. These will be either points of information you want the audience to understand during an informative presentation or items of support for a persuasive presentation. Make sure that each point is clearly separate, that each point is similar in scope or level of importance, that they cover your topic, and that they do not overlap. For example, if you were planning a descriptive presentation about my Zenith personal computer, your main points might be (1) the CPU, (2) the keyboard, and (3) the display unit. You would not include as a fourth point the electric plug unless you wanted to make a humorous point about the importance of the power supply.

Once you have determined the main points you wish to make, you will have to arrange them. Just as in written communication, this arrangement has to make some sort of sense to the audience. Its primary purpose, therefore, is to help the audience understand.

You might choose a chronological pattern of arrangement if your topic has something important to do with time; for example, a procedural topic. You might choose a spatial arrangement if you are describing something physically. Usually this would be augmented with visuals.

Your topic might suggest an associational order of main points, such as cause and effect or problem-solution. Or it might require a multi-part order such as induction, deduction, familiar to unfamiliar, or simple to complex.

If the presentation is persuasive, you might opt for what is known as Monroe's Motivated Sequence. The first part of such a presentation attracts the attention of the audience and then shows the audience that they have a need, or a problem that needs a solution. The presentation then moves on to supply the solution. Next, it provides a visualization of how it would be if the solution were implemented. Then it finishes by outlining how to implement the solution. This is a version of selling, convincing the people of the audience that they need something and then showing them how to get it. But it is a powerful organizational strategy when the presentation purpose calls for it and when it is done right.

Although this section deals primarily with main points in a presentation, if the topic is broad enough and if you have enough time for your presentation, you can subdivide the main points. Only be certain that all the subpoints adhere to the criteria used in selecting and organizing the main points.

OUTLINING THE PRESENTATION

When outlining a presentation, it is important to follow a few widely used guidelines. First, make sure that each level of the outline is similar in scope. This applies to subpoints as well as to main points. Limit each section of the outline to one idea, using a short phrase to remind you of the point you want to make. This makes the outline easier for you to follow if you use it as the notes for your presentation. Make sure that each section does not overlap, so that the audience is not confused. And if you subdivide a main point, there should be two subpoints created by the division. Otherwise, it is like taking an orange, cutting it in half, and having a whole orange left—only smaller than the original.

Focus on transitions in the outline. They are what enables the audience to follow your train of thought. Realize that introductions are transitions. They connect the audience's not knowing about your specific topic to a readiness to be told the specifics of it. Transitions also should tie main points and subpoints together. Don't overlook conclusions. They are what brings a presentation to a successful completion. Each of us has been to a presentation in which the speaker stopped and the audience did not know whether the presentation was finished or not. Such presentations do not conclude; they just end. And even in the informal presentations made within your organization, avoid "that's-about-it" conclusions; those are not conclusions, they're surrenders.

TYPES OF DELIVERY

There are four possible ways to deliver a presentation, each with its advantages and disadvantages. Choose a method that makes you comfortable, but make certain it produces an effective presentation.

Manuscript Method

The manuscript method is what is often referred to as "reading a paper." It is most frequently seen in two situations: large formal conferences that place rigid restrictions on the presentation's length and presentations in which the speaker is terrified to look at the audience.

One advantage to reading a manuscript is that all the speaker's words are determined beforehand. This reduces or eliminates the concern for getting lost or forgetting what one plans to say. It is advantageous for rigid time limits because the entire presentation can be practiced *exactly* as it will be delivered and timed to fit the requirement.

There are disadvantages, as well. Most of us do not read out loud well without training and lots of practice. We tend to read in a monotone, with unnatural rhythms and hesitations. Reading a manuscript also precludes much eye contact with the audience, allowing the audience's attention to wander. If you are not paying attention to them, why should they pay attention to you? To some degree, you can get around this particular problem by writing notes in the margin to yourself saying, "LOOK UP!!!" Then, when you look up, place your thumb on the line you were reading so you can find your way back without tipping off the audience that you are lost.

Generally speaking, limit your use of the manuscript method to formal presentations that are required to be short. Remember that an eight-page, double-spaced manuscript takes almost fifteen minutes to read; much more reading than that wears thin on the audience. Whatever your presentation is, do not resort to the manuscript method as a crutch. The disadvantages of the method are obvious enough to an audience without showing them your fear also.

Memory Method

If you were so inclined, you could memorize a presentation. The advantages are exactly the same as they were for the manuscript method. In addition, memorization is also good for eye contact with the audience.

But the disadvantages are serious. First, a lot of time is required to memorize eight pages of text for a typical 15-minute presentation. And if our reading aloud is bad, our recitation is worse—at least for those of us who lack formal training in this sort of thing. The reason for this is that we memorize not by commiting a string of words to memory but by learning the rhythm patterns inherent in speech and hanging the words on those patterns. That is why a third-grade school play is delivered in the inevitable sing-song; it's how the lines were memorized: "ta-da, ta-da, ta-da, ta-da." To an extent, we adults are guilty of the same thing when we memorize a presentation. And the more we practice it, the worse it gets.

That's not the most serious disadvantage, however. The real danger is memory loss—forgetting what you were going to say, who you are, why you're there. If that happens in a memorized presentation, you might as well sit down.

For these reasons, I cannot think of a single good excuse for memorizing a presentation.

Impromptu Method

This is the easiest way to prepare a presentation: don't prepare at all. Wing it, make it up as you go along, fake it. There are no advantages to this method, at least in terms of advancing your career. If you want to be fired from a company, however, the impromptu method of delivering a presentation just might do it for you.

But there is one unavoidable reason for making an impromptu presentation. You find yourself seated in the audience at a conference; the speaker to whom you are listening notices you are there and remembers you have done work in the area being discussed. Since the speaker has run out of intelligent things to say, he (and it usually is a he, because women tend to treat each other better than this) calls on you to stand up and share your knowledge with the group. At this point, trapped, you slowly get up, blush, stammer out a few unintelligent remarks, proving you know absolutely nothing about the topic or that you weren't paying attention, and sit down, vowing to get even with the speaker if it takes you the rest of your life.

The best thing that can be said about the impromptu method is, "Don't use it."

Extemporaneous Method

This method is preferred by most speakers. It means making a presentation from an outline or notes. It has all the advantages of good eye contact, naturalness of language, rhythm, pace, and voice modulation. It does require practice because it is easy to get caught up in the occasion of conversing with an audience and talk too long or wander off the subject. These disadvantages notwithstanding, you should get in the habit of using the extemporaneous method of presentation delivery most, if not all of, the time.

PRACTICING THE PRESENTATION

Practice is essential for success. Your objective should be a conversational delivery, and the practice should stimulate the actual presentation. This means you should:

- practice out loud
- practice the entire presentation
- practice in the actual presentation environment, if possible, or in one that is similar
- practice using visuals
- practice movement and voice modulation

A participant in a seminar several years ago gave a presentation that demonstrated the importance of practice, especially the right kind of practice. In the interest of time and fairness to all participants, I had imposed a ten-minute limit for presentations. This speaker practiced long and diligently. But during his practice, when he reached a point he was particularly comfortable with, he would say "and so on and so on." When he made the talk, he of course had to fill in the "and so on's." When I stopped him at ten minutes, he was a little over half finished with what would have been almost a twenty-minute presentation.

VOCAL AND BODY DIMENSIONS OF A PRESENTATION

There are many aspects of actually delivering a presentation that are beyond the scope of this book, but nonetheless important. This section will summarize a few of them, but you should consult the books at the end of the chapter for additional information.

Stress of the speaker's voice is important because it can change the meaning of what is being said as well as call attention to particular points in the presentation. Consider the following five sentences as an example. In each sentence, the stressed word is italicized.

(1) The cat is in the washing machine.

(No stress; simple statement of an observation)

(2) The *cat* is in the washing machine.

(As opposed to the gerbil)

(3) The cat *is* in the washing machine.

(As if you did not believe me the first time)

(4) The cat is *in* the washing machine.

(As opposed to being on top of it)

(5) The cat is in the *washing* machine.

(As opposed to being in the dryer)

In each sentence, because of the stress alone, the meaning changes.

Volume is also important. The speaker must be capable of being heard by everyone in the room. This is one more reason why examining the presentation environment before making a presentation is important. If you have a light voice or poor projection, you will want to consider an amplification system (and practice using it).

Enunciation is important. The audience must be able to differentiate among your words. If you mumble or runwordstogether, they will not be able to. This is another reason why the manuscript method should be avoided if possible. When reading a manuscript, you spend a majority of your time speaking to the podium.

You should also pay attention to the pace of your presentation. Realize that if you are anxious at all, you will automatically speak more quickly than normal. If things get completely out of control, your audience will hear a breathless, rushed presentation that is distracting solely by how it is delivered. Try to intersperse silent pauses in your delivery—not with "uhs," "ahs," and worse "ya-know?s"—but with momentary catch-breath silences in which you can collect your thoughts and the audience can digest information before you both proceed. At first when you try this, you will inevitably feel self-conscious, but with practice you can cure the "filler" habit. If you want a model for how this is done well, observe a national network anchor reading the evening news. The silent pauses are natural, unobtrusive, but obviously there.

AVOIDING STAGE FRIGHT

Some degree of anxiety is a reality to all speakers. In fact, it may even be good for you; it gets adrenaline flowing and makes you seem excited about your message. But the important issue is to control it, as well as to remember that audiences are not usually aware of a speaker's anxiety.

If stage fright paralyzes you with fear, you can try to develop relaxation techniques. When you stand up to make a presentation, force yourself to look at the audience—not as a crowd but as individuals. Take a few slow breaths. This does not mean breathe so deeply you hyperventilate, but try to control your breathing and pulse as you start a presentation. Your audience will see what you are doing as getting organized before taking them into your material.

Adequate preparation is also an effective cure for stage fright. The more presentations you deliver, the more confidence you will have doing them.

CONCLUSION

In this chapter, we have examined methods for planning and delivering professional presentations. Practice is essential for success, just as is a consistent focus on communication rather than performance.

But why is making presentations important? One reason is that they are a form of ritual; they provide beginning professionals with opportunities to demonstrate that they "belong to the club." In some ways, presentations are an important part of the corporate culture, enabling speakers to showcase their abilities and reap the rewards of career advancement. Recent research suggests that speakers enter into implicit contracts with their profession and with the audience. In other words, by accepting an opportunity to speak to a group of our colleagues, we are agreeing to make detailed preparations that will support and demonstrate the value of our information, as well as agreeing to present that information in the best way possible. Many of the skills presented in this chapter lead to types of presentations covered in the next chapter, "How to Run Effective Meetings."

SUGGESTED READINGS

Allen, Dave. *In Your Own Voice: A Manual for Public Speaking.* American Press, 1989.

Callahan, Roger. *How Executives Overcome the Fear of Public Speaking and Other Phobias.* Homewood, IL: Dow Jones-Irwin, 1990.

Fletcher, Leon. *Speaking to Succeed in Business Industry Professions.* Harper-Row, 1988.

Frank, Ted and David Ray. *Basic Business and Professional Speech Communication.* Englewood Cliffs, NJ: Prentice-Hall, 1979.

Nelson, Robert B. *Making Effective Presentations.* Scott-Foresman, 1989.

Tacey, William S. *Business and Professional Speaking.* Dubuque, IA: William C. Brown Company, 1980.

CHAPTER 22

How to Run Effective Meetings

In all businesses, an important amount of work is carried on in meetings through small-group discussions. Often, those of us who have to sit through a lot of these become convinced that there has to be a more efficient way of running things. Too many meetings seem dedicated to the inverse law of meeting management: the longer the meeting, the less that gets done. Fortunately, it does not have to be this way. Successful meetings depend on the following factors:

- leadership ability
- problem-solving skills
- listening skills

Unfortunately, these skills are often thrown down and trampled on during most business meetings. This chapter will show you ways to run effective meetings. Figure 21 is an outline of how to do this.

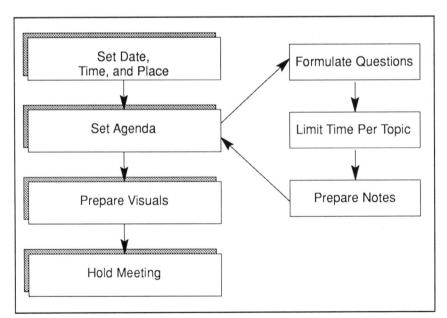

Figure 21. VOS for Meetings

Remember, however, that in small-group discussions perhaps more than in any other type of communication, personality differences come into play. Different people react in different ways, and since in industry most writing projects are the result of collaborative effort (interpreted as plenty of meetings), it would probably be a good idea to review the material in Chapter 1 to develop the interpersonal skills for handling these differences and for turning them into strengths.

LEADERSHIP ABILITY

All groups have leaders. For planned meetings, the role of leader is generally assigned. In other words, if a manager calls a meeting, everyone knows who the leader is. With informal discussions, the situation is different and less rigid, but a leader will emerge.

Regardless of the meeting's purpose or environment, leaders do similar things. They can set the tone of the meeting, making it serious or easy-going. They order the flow of business by manipulating the discussion. Good leaders do this for the purpose of solving group problems or covering the topic to be discussed. Some leaders, however, can do the same thing with only self-serving intentions, playing up their own sense of importance. For this reason, part of a meeting's success depends upon the leader's integrity. Regardless of leaders' intentions, they order the flow of information in a meeting by interjecting questions and comments during a discussion. In very formal meetings, this is enhanced by a rigid agenda with time allotments for each topic to be considered.

Leaders also listen. They hear agreement among members of the group, aid exploration of what is agreed upon, and help make decisions.

PROBLEM-SOLVING SKILLS

Discussions that solve problems depend upon a willingness of all members of the group to communicate their ideas freely, to cooperate with others, to listen to their ideas, and to give and receive constructive criticism of ideas. This results in group members contributing their knowledge to the problem-solving process. Every member of a group has something valuable to say, if for no other reason than to present a different viewpoint on a problem. Being able to criticize the ideas that are discussed and being able to receive criticism of your own ideas is the only way to ensure this free interchange of information. Finally, cooperation among members of the group makes it likely that something useful will come out of the discussion.

To facilitate problem-solving within meetings, leaders should attempt the following:

Ask a question. Asking a question as a way of organizing discussion is preferred over choosing a topic because it focuses the group's attention on the problem. Questions have answers; topics only have subtopics. Choosing a topic or presenting a topic often leads to the seemingly endless and almost always fruitless discussions which all of us have experienced.

Define a question. This requires the group to agree on precisely what is at issue. Doing this builds a fence around what is to be discussed and enables leaders to keep the discussion centered on the problem.

Examine the background. The only way problems can be solved is if the group understands them. Part of this understanding is knowing what led to the problem in the first place. This is an extension of defining the question.

Select criteria for evaluating suggested solutions. This is another fence-building task. It enables the group to agree on what should be important in proposed solutions. Then if some solutions do not meet these criteria at the outset of the discussion, they can be discarded.

Suggest solutions. In successful problem-solving meetings, a number of alternatives will be examined.

Discard all but the best solution. This is still the most time-consuming task of group problem solving. Even in the best of situations, agreement will rarely be easy to come by. But following this procedure will at least ensure that the meetings you lead will arrive at this stage as efficiently as possible. The procedure enables groups to avoid wandering down unproductive tangents.

LISTENING SKILLS

Effective listening requires that meeting leaders let others talk. Limit responses and questions to the main point of discussion. This makes sure that you are not overbearingly imposing your will on the direction of the group, but it also allows you as a leader to keep the discussion focused.

Attempt to clarify what other members of the group have said. This summarizing allows time for the group to reflect on what has been discussed, time to gather thoughts and evaluate. Resist telling the group what they should be thinking and saying. This is hard for some types of people to do as leaders, but if you can master it, your leadership and problem-solving will be better as a result.

Try to understand solutions and ideas from the other group members' perspectives. This demonstrates an openness, a willingness to discuss alternatives. It helps involve all group members in the problem-solving process. This is true even if the final decision is yours to make.

Pay attention to courtesy. All of us have experienced situations in which we knew what the other person was going to say or how an idea was going to be developed. The temptation, particularly among people who are intuitives, is to interrupt and finish the other person's thought. *Don't do it!* Not only is this rude and disrespectful of the other person, it destroys the cooperation of the group. Do too much of this sort of thing, and other group members will quit speaking freely.

CONCLUSION

This chapter has presented some of the techniques for leading effective meetings. It is not a complete discussion of this large task. Completeness would require a separate book, a book that I have included in the suggested readings at the end of this chapter. Nonetheless, leading meetings and small-group discussions is a vitally important aspect of communicating within the high-tech industries. Brain-storming design sessions are common. And more and more, much writing is done collaboratively. Problems are identified in meetings, solutions discussed, documents reviewed. Too often, these can be self-promoting affairs because of how strongly egos are wrapped up in the design or writing process. Subordinating one's ego to the group is not a denial of integrity. Rather, it is a realization that effective group thinking makes meetings more effective and *shorter*.

SUGGESTED READINGS

Athos, Anthony G., and John J. Gabarro. *Interpersonal Behavior*. Englewood Cliffs, NJ: Prentice-Hall, 1978.
Haney, William V. *Communication and Interpersonal Relationships*. Homewood, IL: Irwin Press, 1979.
3-M Meeting Management Team. *How to Run Better Business Meetings*. New York: McGraw-Hill, 1987.

CHAPTER 23

How to Use Visuals with Presentations

Visuals are an important addition to presentations. But that is all they are; they cannot and should not take the place of language in lengthy communications. Rather, they should be used to highlight presentations, to focus the audience's attention on important material. Correctly designed and used, visuals should support and expand the content of your presentation. They should clarify what you mean as you speak. If they are to do this, they must meet certain criteria:

- They must be visible, large enough for the *whole* audience to see— even those people who insist on sitting in the back row.
- They must be clear; their meaning must be obvious at a glance *without explanation*.
- They must be simple and easy for the audience to comprehend.
- They must be controllable, easy for you to use with your presentation.

By now you are no doubt thinking that you have never seen visuals that met all these criteria, particularly the one about their meaning being obvious at a glance without explanation. How many speakers have we listened to who have spent the majority of their time explaining their slides rather than giving us the information we came to receive? However, the fact that the majority of presentations fall short of these standards is not a justification for ours to fall short of excellence, as well. Good visuals polish a presentation. For that reason alone, good visuals should be our goal.

There are many types of visuals available for the speaker to use, each with advantages and disadvantages. In this chapter, the different types of visuals will be examined with special attention paid to the type of presentation environment in which they are best used.

WORDS AND PHRASES

Words and phrases are points upon which the speaker will elaborate. They should represent important matters in the presentation, and they should be short and simple. Frequently, speakers will present the audience with an outline of their presentation at the beginning of it. Although this treatment is a little heavy handed, it does work in that the audience is automatically oriented to the topic and to the order in which it will be treated.

Words and phrases might be presented on charts, slides, view graphs, or chalkboards with equal success as long as the criteria for the use of each of these media is also met. Just make sure that you do not force listeners to do too much reading; if they do, they won't be listening to you.

CARTOONS

Cartoons, or exaggerated figures, are effective visuals when the speaker is dealing with a sensitive subject. This would be a fairly rare occurrence in the high-tech industries, but cartoons can also be used to depict people-oriented action in a stationary medium or to enliven less-than-exciting material. These two uses are often useful when making a presentation about high technology—especially to an audience that does not share the speaker's knowledge of the subject.

CHALKBOARDS AND MARKER BOARDS

Chalkboards and marker boards are popular media for informal, small-group presentations. In spite of their popularity, most people who use them misuse them. If you are planning to depict information on a chalkboard or marker board during a presentation, write the information on the board *before* the presentation and cover it. Then you can reveal the information at the proper time during your presentation. Make sure that what you write on the board is simple and neat. Successful use of these boards requires a little practice; it isn't as easy as it looks. The first time you do it, you will notice how odd writing at a vertical angle is and how difficult it is to write legibly.

Once you are ready to reveal the information to the audience, prime them. Let them know what is coming up and why it is important *before* you show it to them. Remember also that chalkboards and marker boards should be limited to fairly small presentation environments. People who are seated more than 30-40 feet from the board will have a hard time reading what you have written. And if you write very large, you will use up so much of the board that it will not be an effective visual medium. Remember, too, that if you are using a marker board, make sure you have a dry marker to use instead of a magic marker. If you don't, you will create more problems for yourself than you can imagine: you can't erase magic marker.

Finally, whatever you do, avoid turning your back on the audience, writing on a board, and talking to what you have written. This is a guaranteed way to distract an audience.

CHARTS

There are two types of charts that speakers may use—hardboard charts and flip charts. Hardboard charts can be placed on an easel and revealed one at a time; flip charts can be placed on a support with hooks and flipped (or torn off) page by page. Both types of charts are useful for moderately large audiences—up to 100 people seated no more than 75 feet from the charts. Realize, however, that what you can do with charts varies widely depending on whether the audience is small and located in a conference room or large and seated in an auditorium. With a large audience, about all you can present on charts is short words and phrases, printed very large. But with small audiences, figures, graphs, and tables are also effective, provided that the audience can read *all* the information.

That last statement calls to mind criteria that charts must meet to be effective. First, the material presented must be simple. Detailed work—even on a large chart—is almost impossible for the audience to see. Second, when using hardboard charts, it is a good idea to have an assistant who can help remove the charts. Third, just as was the case with chalkboards and marker boards, charts should be predrawn—even flip charts. And fourth, the audience should be primed for the information that will be presented to them.

VIEW GRAPHS OR TRANSPARENCIES

View graphs, or overhead transparencies, are useful for presenting information to relatively large audiences. By adjusting the projection distance, the images can be made quite large without a loss of detail if the projection equipment is good.

View graphs can be used in a variety of ways. You can overlay them to add on information. You can cover part of them and reveal information as it becomes appropriate to your topic. You can even write directly on the transparency, although this is the least effective and most distracting use. You should not, however, place a page of typed text on the view graph and expect your audience to read it. They won't.

An important disadvantage of these type of visuals, however, is the fact that they can be difficult to manipulate. They are prone to static electricity, so they stick together worse than wool socks in a clothes dryer. If you are the slightest bit anxious during your presentation and you have to contend with stuck transparencies, you will likely break out into hives. To combat stuck transparencies, fan them out rather than placing them in a neat stack.

SLIDES

Slides are excellent for presenting information to large audiences. They make it possible to organize the entire presentation visually, store it in a carousel, and use it over and over again. Remote control is another advantage in that it allows you the freedom to move about in your presentation. The other types of visuals considered so far limit your movement.

Effective slides meet a different set of criteria. First, if you use them in a completely darkened room, be sure that you light yourself at the podium. Some people find it difficult to pay attention to a disembodied voice in the darkness; they are likely to take a nap. Second, break the presentation into segments of six to ten slides with additional commentary between each segment. This helps to combat any rigidity or tedium a slide presentation might have. Third (once again!), prime the audience. Let them know what each slide means, what each segment of slides means. Tell them why each slide is important.

MOVIES AND VIDEOTAPES

Movies and videotapes are possibly the most versatile type of visual available. They can even do away with your role as speaker. Unfortunately, this is often the tendency. The correct use of movies and videotapes requires that you again prime the audience, that you talk about the videotape or movie, explaining its relevance to the topic. If you use these visuals to your advantage, they can be very effective, stimulating questions and discussion.

COMPUTER SCREEN PROJECTIONS

Since the first edition of this book was published, I have seen more and more use of computer screen projections as a visual technique in professional presentations. And to be honest, most have not worked very well. Although it will likely change (and no doubt soon), the technology is not yet at a stage where computer screen projections are crisp and clear. In addition, watching the speaker/user type entries or manipulate an arrow-pointer on the projection is distracting. So, are computer screen projections a case of a technology that has not yet matured to its proper place and potential use, or is it a case of more technological power than makes sense for clear communication purposes? It's too soon to tell.

USING VISUALS

Using visuals takes practice. While each type of visual has advantages and disadvantages for presenting information, each also has its problems for the speaker in terms of using it "on the spot."

Chalkboards and marker boards are probably the easiest visual for speakers to use once the problem of becoming comfortable writing on them is solved. No doubt this is why they are popular.

Charts, particularly hardboard charts, have a tendency to fall on the floor if there is even the slightest hint of a breeze. Using clips can help eliminate this problem. With flip charts, the simple act of flipping them or tearing them off is distracting. Speakers have to practice repeatedly with charts to use them smoothly and avoid unnecessarily distracting audiences.

Some of the problems using view graphs have already been mentioned. Add to these the frustration of dealing with them should you drop them on the floor and the frustration to the audience if you keep putting them on the projector upside-down or backward. Furthermore, most of us know of at least one person who has been using a projector when the bulb burned out. Always carry a spare!

Slides can be a problem if the speaker has not run through the carousel to check whether all the slides are in order and right-side-up. Some speakers also have a tendency to go berserk with the remote control device, flipping slides forward too far and then overcompensating by flipping back too far.

Novice movie projectionists run the risk of tearing the film, provided they have been able to get it threaded in the first place. Videotape playback machinery is relatively trouble free, if the speaker has taken the time to learn how to operate it. And computer screen projections have problems that are yet to be discovered, not the least of which is potential eye fatigue for audience members.

CONCLUSION

There are no perfect visuals. Each type has advantages and disadvantages, both in their ability to present information and in our ability to use them. In this chapter, we have examined the different types of visuals available to us, looking at how they can add to our presentations and how they can take away from them if we are not comfortable with their use. That last point is the most important: choose visuals that will add to your presentation, that will fit into the presentation environment you will be using, and that you feel comfortable with. Then *practice, practice, practice*!

Index

by Linda Webster

CHARLES H. SIDES is technical communication coordinator, Fitchburg State College, Fitchburg, MA. He is the author of many works on technical writing.